Federated Learning for I of Vehicles: IoV Image Processing, Vision and Intelligent Systems

(Volume 1)

Reinventing Technological Innovations with Artificial Intelligence

Edited by

Adarsh Garg

G.L. Bajaj Institute of Management and Research, Greater Noida, India

Valentina Emilia Balas

Automatics and Computer Science
Aurel Vlaicu University of Arad
Arad, Romania

Rudra Pratap Ojha

G. L. Bajaj Institute of Technology and Management, Greater Noida, India

&

Pramod Kumar Srivastava

Rajkiya Engineering College
Ajamgarh, Uttar Pradesh, India

Federated Learning for Internet of Vehicles: IoV Image Processing Vision and Intelligent Systems

(Volume 1)

Reinventing Technological Innovations with Artificial Intelligence

Editors: Adarsh Garg, Valentina Emilia Balas, Rudra Pratap Ojha & Pramod Kumar Srivastava

ISBN (Online): 978-981-5165-79-1

ISBN (Print): 978-981-5165-80-7

ISBN (Paperback): 978-981-5165-81-4

need for a court order if at any point you breach any terms of this License Agreement. In no event will any delay or failure by Bentham Science Publishers in enforcing your compliance with this License Agreement constitute a waiver of any of its rights.

3. You acknowledge that you have read this License Agreement, and agree to be bound by its terms and conditions. To the extent that any other terms and conditions presented on any website of Bentham Science Publishers conflict with, or are inconsistent with, the terms and conditions set out in this License Agreement, you acknowledge that the terms and conditions set out in this License Agreement shall prevail.

Bentham Science Publishers Pte. Ltd.
80 Robinson Road #02-00
Singapore 068898
Singapore
Email: subscriptions@benthamscience.net

BENTHAM SCIENCE

CONTENTS

Adarsh Garg and *Amrita Jain*

PREFACE

Augmented and Virtual Reality (AVR) technology shapes the margins between the real world and the digital world. The boundary between the two worlds has become so porous that it nudges into more personalized and unique experiences across a range of industries like tourism, marketing, education, social media, construction, and so on. However, its more revolutionary impact would be to the extent where each one of us would be capable of transforming our real surroundings into digitally enhanced personal experiences with a very thin margin between the real and digital worlds. This is going to transform the whole business world and further revolutionize our surroundings *i.e* Society 5.0.

AR and VR are widely considered, so far, in the gaming and entertainment sectors. But looking at the Gartner report on strategic technology trends 2019, AR and VR technologies are among the top 10 trends and it has been experimented to enhance the productivity of various business domains like marketing, construction, power, education, tourism, telecommunication, automobiles, sports, *etc*. AR technology joins the information and simulated images in the real milieu to augment the users' circumstantial perception of their settings, using some technology-equipped devices whereas VR technology replaces the users' perception of their surroundings to complete virtual surroundings with the help of computers only. The augmentation might be the result of AR, VR or both AVR. Construction industries are using AVR to reduce the risks at working sites. AVR technologies are of extreme significance for the construction segments as the assembled setting is essentially connected to a 3D space and AEC professionals rely mainly on imagery for communication. Boeing's aerospace giant has been using AVR for electricals. A very famous company of UPS is using AVR (AR & VR separately) to provide training for driver safety. However, AVR technologies have not yet been explored, or used to an extent that would make them more reliable for realistic business necessities. Technical limitations, lack of awareness, resistance to use and accept AVR as a substitute, high cost, and time obligation are some of the major challenges of bursting usage of AVR.

Irrespective of the AVR maturity level, there is a need to focus on the effectiveness of AVR technologies to enhance innovations, and sustainability in various business domains, particularly construction tasks, and for the targeted implementation of research. There is no book volume as of now that focuses on the entirety of AVR's strengths, its reliable usage in business innovations and the challenges to be addressed by the researchers. The proposed book volume will address all these points with a more holistic approach, ranging from awareness to innovations and reliability to sustainability from a business perspective. More specifically, no edited volume exists that systematically maps (i) how AR and VR technologies can be used, (ii) their potential benefits, (iii) prevalent issues, and (iv) a futuristic innovation plan.

The book on AI innovations is organized as follows.

Chapter 1 throws light on the use of multi-agent systems which often operate in dynamic, open, and complicated settings. Two approaches to improving agent interactions are presented in this chapter. By using ontologies, the technique may allow agents to create "rich" interaction protocols using Petri net (CPN) based methodologies in order to allow agents to create dynamic protocols.

Chapter 2 describes the need for Artificial intelligence (AI) which has gained enormous usage in business in recent years. But the use of Artificial Intelligence is limited to a greater extent when it comes to measuring business ethics and morality. The chapter, conceptually formulates the implementation of AI in CSR programs by using AMOS 21's Structural Equation Modelling (SEM) and SPSS 21 with empirical testing of projected models for AI efficient CSR practices.

Chapter 3 emphasizes how AI integrated with machine learning (ML) and Deep-learning (DL) techniques are used in various disease diagnosis domains, medication discovery, medical visualization, digital health records, and electro-medical equipment.

Chapter 4 discusses the method of combining information in the form of image alternatives with a software programme that stores knowledge with real images. Augmented and virtual reality (AVR) technologies aid in the explanation of concepts to improve academic learning through the use of two-dimensional media in education.

Chapter 5 discusses the role of VR in 3D reconstruction and visualization of crime situations such as criminal assaults, traffic accidents, and homicides by establishing a new method for criminal investigation.

Chapter 6 explains how rapid advances in artificial intelligence are enhancing the performance of many sectors and enterprises, including green supply chain management. It further analyzes the future outlook of the market for Artificial Intelligence (AI) in GSCM and green sustainability if they follow SDGs.

Chapter 7 discusses the use of information-driven systems to offer problem-specific knowledge to decision-makers using internet-based distributed platforms. An XML-based approach to representing and exchanging domain-specific information for informed decision support is shown in the chapter. The technology's implementation specifics, commercial ramifications, and future research goals are presented.

Chapter 8 portrays the importance of the farming sector which is considered to be the backbone of the Indian economy. The work emphasizes on the use of an automated watering system to reduce the farmer's manual involvement in the field at an effective cost by implementing an artificial intelligence system based on sensing, a control mechanism with required correction for the maximum yielding of irrigation.

Chapter 9 introduces AI as a useful aid to urban planning thereby creating a safer and more sustainable future for its citizens. Applications of AI in smart cities are then discussed, followed by a brief discussion on the prevailing best practices. Challenges in creating AI-enabled smart cities in India are also outlined in the chapter.

Chapter 10 portrays that Augmented Reality is the need of the hour for Human Resource Management in this era of globalization wherein the world has become flat and businesses have no boundaries. The chapter presents the evolution, applications, and challenges of VR and AR with respect to HRM.

Chapter 11 intends to explore how AI-enabled technology, in the fashion industry and fashion environment, is influencing the green economy status of the fashion industry, especially in the post-COVID-19 era of innovative e-commerce fashion.

The work given in the book will give some interesting insights to the readers.

Adarsh Garg
G.L. Bajaj Institute of Management and Research
Greater Noida
India

Valentina Emilia Balas
Automatics and Computer Science
Aurel Vlaicu University of Arad
Arad, Romania

Rudra Pratap Ojha
G. L. Bajaj Institute of Technology and Management
Greater Noida
India

Pramod Kumar Srivastava
Rajkiya Engineering College
Ajamgarh
Uttar Pradesh, India

List of Contributors

Ankita Tiwari	Department of Engineering Mathematics, Koneru Lakshmaiah Education Foundation, Vaddeswaram, AP, India
A. Menaga	School of Management Studies, Vels Institute of Science, Technology & Advanced Studies(VISTAS), Chennai, India
Amit Bhaskar	Rajkiya Engineering College, Azamgarh, Deogaon, Azamgarh, Uttar Pradesh, India
Akriti Dutt	Agriculture Department, Government of Uttar Pradesh, Uttar Pradesh, India
Aditya Saini	Forensic Science, School of Basic and Applied Sciences, Galgotias University , Greater Noida, Uttar Pradesh 203201, India
Archana Singh	Faculty of Commerce & Management, Vishwakarma University, Pune, Maharashtra 411048, India
Adarsh Garg	Data Analytics, G. L. Bajaj Institute of management and Research, Greater Noida, India
Amrita Jain	Data Analytics, G.L. Bajaj Institute of management and Research, Greater Noida, India
Brihaspati Singh	Rajkiya Engineering College, Azamgarh, Deogaon, Azamgarh, Uttar Pradesh, India
Divya Pratap Singh	Department of Applied Sciences and Humanities, Rajkiya Engineering College, Azamgarh, Uttar Pradesh, India
Jagjit Singh Dhatterwal	Department of Artificial Intelligence & Data Science, Koneru Lakshmaiah Education Foundation, Vaddeswaram, AP, India
Kajol Bhati	Forensic Science, School of Basic and Applied Sciences, Galgotias University , Greater Noida, Uttar Pradesh 203201, India
Kabaly P. Subramanian	Faculty of Business Studies, Arab Open University, Halban, Oman
Kuldeep Singh Kaswan	School of Computing Science & Engineering, Galgotias University, Greater Noida, India
Manisha Singh	Economics and Strategy, G.L. Bajaj Institute of Management and Research, Greater Noida, India
Narendranath Uppala	Putra Intelek International College, Petaling Jaya, Malaysia
Neerja Aswale	Faculty of Commerce & Management, Vishwakarma University, Pune, Maharashtra 411048, India
Pankaj Yadav	Rajkiya Engineering College, Azamgarh, Deogaon, Azamgarh, Uttar Pradesh, India
Pooja Agarwal	Faculty of Commerce & Management, Vishwakarma University, Pune, Maharashtra 411048, India
Rajeev Kumar	Forensic Science, School of Basic and Applied Sciences, Galgotias University , Greater Noida, Uttar Pradesh 203201, India
Radheshyam Dwivedi	Department of Electrical Engineering, MNNIT Allahabad, UP, India

S. Vasantha	School of Management Studies, Vels Institute of Science, Technology & Advanced Studies(VISTAS), Chennai, India
Savendra Pratap Singh	Rajkiya Engineering College, Azamgarh, Deogaon, Azamgarh, Uttar Pradesh, India
Sambhrant Srivastava	Rajkiya Engineering College, Azamgarh, Deogaon, Azamgarh, Uttar Pradesh, India
Saurabh Kumar Singh	Rajkiya Engineering College, Azamgarh, Deogaon, Azamgarh, Uttar Pradesh, India
S. Christina Sheela	Gnanam School Of Business , Sengipatti, Tamil Nadu 613402, India
S.P.S. Arul Doss	Gnanam School Of Business, Sengipatti, Tamil Nadu 613402, India
Shyam Narayan Singh	Forensic Science, School of Basic and Applied Sciences, Galgotias University , Greater Noida, Uttar Pradesh 203201, India
S. Susithra	School of Management Studies, Vels Institute of Science, Technology & Advanced Studies(VISTAS), Chennai, India
Vijay Kumar	Rajkiya Engineering College, Azamgarh, Deogaon, Azamgarh, Uttar Pradesh, India
V. Selvalakshmi	Srm Valliammai Engineering College , Kattankulathur, Tamil Nadu 603203, India
Vinny Sharma	Forensic Science, School of Basic and Applied Sciences, Galgotias University , Greater Noida, Uttar Pradesh 203201, India
Vibhooti Narayan Mishra	Department of Mechanical Engineering, NIT Patna, Bihar, India
Yasmeen Bano	School of Management Studies ,Sathyabama Institute of Science and Technology (SIST), Chennai, India

List of Abbreviations

AAL	Ambient Assisted Living applications
ACLs	Agent Communication Languages
AGFI	Adjusted Goodness of Fit Index
AGI	Artificial General Intelligence
AI	Artificial Intelligence
AIBO	Series of Robotic Dogs
ALS	Alternate Light Sources
ANN	Artificial Neural Networks
AR	Augmented Reality
ASI	Artificial Super Intelligence
ATS	Agent Type Set
AUC	Area Under the ROC Curve
AVR	Augmented and Virtual Reality
BI	Business Intelligence
CAD	Computer Aided Design
CAGR	Compound Annual Growth Rate
CARE	Center of Alternate & Renewable Energy
CEO	Chief Executive Officer
CFI	comparative Fix Index
CMIN/Df	Chi-square Fit Statistics/Degree of Freedom
COO	Chief Operating Officer
COVID-19	Coronavirus Disease
CPN	Colored Petri Net
CSI	Corporate Social Irresponsibility
CSR	Corporate Social Responsibility
CVA	Computer Vision Algorithms
DA	Discriminant Analysis
DAML+OIL	Ontology Language for Semantic Web
DC	Direct Current
DGM	Deep Generative Models
dhh	Difficult of Hearing
DL	Deep-Learning

DLM Deep-Learning Model

DLM Dynamic Learning Maps

DNA Deoxyribonucleic Acid i

DOM Document Object Model

DSS Decision Support Systems

DT Decision Tree

DTD Document Type Definition

EHRs Electronic Health Records

ELM Extreme Learning Machine

EMR Electronic Medical Records

EMRs Electronic Medical Records

ESG Environmental, Social, and Governance

FIPA's Foundation for Intelligent Physical Agents

GA Genetic Algorithm

GA Google Analytics

GAN Generative Adversarial Networks

GE Green Economy

GFI Goodness-of-fit Index

GNN Graph Neural Networks

GPS Global Positioning System

GPU Graphics Processing Unit

GSCM Green Supply Chain Management

GTS Global Telecommunication System

GVA Gross Value Added

HIV Human Immunodeficiency Virus

HMDs Health Monitoring Devices

HRMD Human Resources Management & Development

HTML HyperText Markup Language

HUD Head-Up Displays

IBM International Business Machines

ICT Information & Communication Technology

IDC International Data Corporation

IDSS Intelligent Decision Support Systems

IMD Institute for Management Development

IMU Inertial Measurement Unit

IoT	Internet of Things
KBS	Knowledge-based Systems
KM	Knowledge Management
KMSs	Knowledge Management Systems
KNN	K-nearest Neighbor
KQML	Knowledge Query and Manipulation Language
KR	Knowledge Repositories, Knowledge Representational
KSL	Knowledge Systems, AI Laboratory
LASSO	Least Absolute Shrinkage and Selection Operator
LCD	Liquid-Crystal Display
LDA	Latent Dirichlet Allocation
LR	Logistic Regression
LUAD	Lung Adenocarcinoma
LUSC	Lung Squamous Cell Carcinoma
MAS	Multi-Agent System
MCDM	MULTI-CRITERIA DECISION-MAKING
MIT	Massachusetts Institute of Technology
ML	Machine Learning
MoHUA	Ministry of Urban Development
NASA	National Aeronautics and Space Administration
NB	Naïve Bayes
NPL	Natural Language Processing
NTS	Non-understandable Type Set
PA	Place Attribute
PACS	Picture Archiving and Communication Systems
PCB	Printed Circuit Board
PHR	Personalized Health Records
PLS-DA	Partial Least-Squares Discriminant Analysis
PMCs	Project Management Consultants
PNs	Petri Nets
PPP	Public Private Participation
PT	Place Type
PTS	Place Type Set
PV	Photovoltaic
PV	Photo-Voltaic

RDF Resource Description Framework

RDF Resource Description Framework

RF Radio Frequency

RF Random Forest

RL Reinforcement Learning

RNN Recurrent Neural Networks

ROC Receiver Operating Characteristic

RPA Robotic Process Automation

RPART Recursive Partitioning

RTS Regional Transit System

SCM Smart Cities Mission

SDG's Sustainable Development Goals

SEM Structural Equation Modelling

SME's Small and Medium Enterprises

SOAP Simple Object Access Protocol

SPSS Statistical Package for the Social Sciences

SPV Special Purpose Vehicle

SSCM Sustainable Supply Chain Management

STAR Smart Tissue Autonomous Robot

SUTD Singapore University for Technology and Design

SVM Support Vector Machine

SVM Support Vector Machines

TVET Technical and Vocational Education Training

UAV Unmanned Aerial Vehicle

ULB Urban Local Body

UNEP United Nations Environment Programme

USB Universal Serial Bus

UTS Understandable Type Sets

VAE Variation Autoencoders

VR Virtual Reality

VRD Virtual Retinal Displays

VRLA Valve-Regulated Lead Acid

VWC Volumetric Soil Moisture

W3C World Wide Web Consortium

WHDs Wearable Health Devices

WSN Wireless Sensor Network

XML EXtensible Markup Language

XR Extended Reality

CHAPTER 1

Agent Interactions Environments

Kuldeep Singh Kaswan[1,*], **Jagjit Singh Dhatterwal**[2] and **Ankita Tiwari**[3]

[1] *School of Computing Science & Engineering, Galgotias University, Greater Noida, India*

[2] *Department of Artificial Intelligence & Data Science, Koneru Lakshmaiah Education Foundation, Vaddeswaram, AP, India*

[3] *Department of Engineering Mathematics, Koneru Lakshmaiah Education Foundation, Vaddeswaram, AP, India*

Abstract: Any system capable of acting as an intelligent agent has all of these characteristics. When an agent has the capacity to interact with other agents, it is able to do so in a multi-agent system (MAS). Systems with several agents often operate in dynamic, open, and complicated settings. Many factors, such as domain restrictions, the number of agents, and the interactions between agents, are not fixed in an open environment. There are several problems in coordinating the interactions and cooperation of agents; as a result of this, many existing agent interaction protocols are not well-suited for open settings, which is a significant impediment to agent interaction. Two approaches to improving agent interactions are presented in this chapter. To begin, by using ontologies, the technique may allow agents to create "rich" interaction protocols. When it comes to agent interaction in open settings, we employ colored Petri net (CPN) based methodologies in order to allow agents to create dynamic protocols.

Keywords: Constraints Function, Agent Communication Language, Agent Interaction Protocols, Conceptual Frameworks, Computational Science, CPN, Intelligent Physical Agents, Multi-agents, MAS Ontology, Standard Protocol, Supervised Learning.

INTRODUCTION

One of today's most essential design ideas is multi-agent systems. Computational systems that include intelligent agents are called multi-agent systems (MAS). If you want to know what's going on in the world around you at any given time, you need an intelligent agent. There are four key characteristics of intelligent agents in general [1]:

---

* **Corresponding author Kuldeep Singh Kaswan:** School of Computing Science & Engineering, Galgotias University, Greater Noida, India; E-mail: kaswankuldeep@gmail.com

Adarsh Garg, Valentina Emilia Balas, Rudra Pratap Ojha & Pramod Kumar Srivastava (Eds.)

- Self-control and the opportunity to interact with and work with other agents is a key aspect of social intelligence, which is characterized by autonomy and self-control.
- Agents' social skills may be honed *via* the use of MASs. There are MAS agents that live and work together in the same family. In a multi-agent society, it is difficult to control the connections between the many actors. When one of the agents chooses to influence others to attain a set of objectives,they get involved with one another. The exchange of messages and declarative interpretations of textual information creates interactions between agents in a system [2].
- Agent communication languages (ACLs) include Knowledge Query and Manipulation Language and the Foundation for Intelligent Physical Agents (FIPA's) ACL (FIPA, 2004).
- Protocols for agent interaction specify common patterns for communications sent back and forth between them. Because of the limitations of many current agent interaction protocols, MASs cannot be used in a broad variety of contexts [3].

As a first step, many current MASs application sectors need agents to operate in dynamic and unexpected (open) settings. Interaction among agents in these situations may be affected by unexpected messages, message loss, or message order abnormalities. Agent-interaction procedures as they now exist are unable to cope with the unforeseen situations that may arise. Secondly, certain MASs have a variety of agent designs, and different agents may interact in different ways [4]. One agent can't be sure that the other agents will comprehend or accept the discussion he or she conducts with the other. To make problems worse, the vast majority of agents have interaction protocols hard-coded into their programming. Agent designers are in charge of determining whether to use a certain protocol, what data to send, and how to carry out tasks in the proper order. Changing protocols after they have been pre-programmed into an agent is a trade-off. KQML, for example, is a modern interaction protocol that isn't specifically designed to transfer knowledge [5]. No one should use this "poor" (Lesser, 1998) method of sharing complex information. Many existing interaction protocols are rigid and inflexible, which make it difficult to implement MASs. In this regard, MASs researchers are working to establish a flexible and knowledge-rich interaction protocol [6].

A technique for agent relationships is covered in this chapter that may enhance both theoretical and practical aspects of agent interactions. Agents may design "knowledge-rich" protocols for interfacing as a first step using this method. An ontology facilitator is a person who helps agents identify, acquire, and develop ontologies [7]. Colored Petri nets (CPNs) may be used to construct a strategy that allows agents to dynamically establish interaction protocols, which indicates that it is not the job of agents to create protocols; instead, agents use their talents and condition to determine what protocols should be used.

Here is a breakdown of the rest of the denomination's structure: Both ontology-based MASs and the usage of PNs and CPNs to specify agent procedures are discussed in this work, which is divided into two sections. In the fourth part, agents may use CPN-based approaches to construct dynamically flexible protocols. To conclude this denomination's methodology section, its is explored for potential applications. The project's results and future intentions are summarized in this section.

ONTOLOGY-BASED INTELLIGENT AGENT INTERACTION

Agents require common terminologies to construct their knowledge and theoretical frameworks of the topic of interest to accomplish knowledge-level communication. A semantic web or a computer language may be used to build ontologies, in which these conceptualizations can be articulated. There must be a common ontology for the MAS's working environment to allow agents to create knowledge "rich" interaction protocols. Ontology facilitators should be included to help agents seek, acquire, and construct conceptual frameworks [8].

Multi-Agent System Ontology Expressions

The intellectual discipline of philosophers is where the term "ontology" comes from. It is possible for an agent or a group of agents in MASs to have an ontology that is computer-readable interpretation of knowledge regarding ideas, connections, and limitations.

MASs ontology

In general, MAS ontologies may be divided into two types: common ontologies and special ontologies. It is possible to create broad ontologies, which explain the aggregate knowledge of an entire multi-agent society, and more narrow conceptual frameworks, which define the understanding of just one particular agent in that society. An ontology representations format and standard working domain ontologies are both necessary components of the MAS design process. Several renowned supervised learning research institutes have already developed standard ontologies for a broad variety of application disciplines as a consequence of the advantages of predictive modeling (for example, the Stanford KSL Ontolingua Server) [9].

As a result, MAS domain ontologies may be created or current ontologies can be referenced.

Ontologies are conventions for machine-readable understanding, and they are commonly expressed in Semantic web technologies such as RDF or computational science which are formal languages. Ontology interpretations still lack a balanced scoring methodology (format). That is, there are a number of ontology languages that have been extensively and effectively employed in a range of application domains. As an example, in several applications, many researchers have found success using DAML+OIL [10]. It showed that ontologies may be used to describe expertise in an online auction mechanism by evaluating the benefits of many commonly used ontology technologies. As seen in Fig. (1), an "item" is used as an example of how one can express an ontology in the digital commerce involved in transportation. OIL is used to represent the ontology in this example.

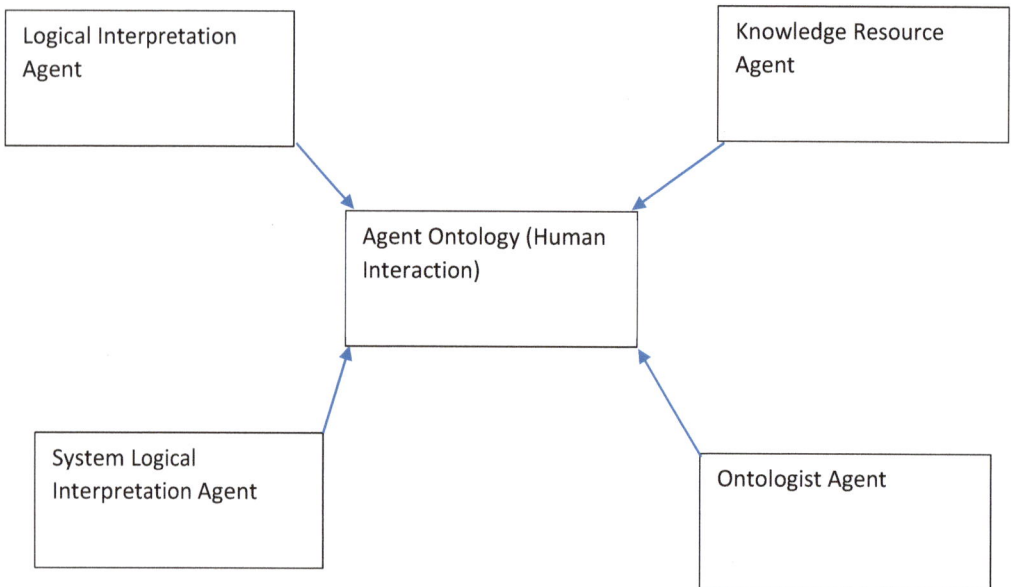

Fig. (1). Ontology Framework.

Editable Ontology-Based MASs

An ontology-based MAS's conceptual framework must contain an ontology facilitator, which makes it easier for agents to find, acquire, and change ontology data. The methodology for ontology-based MASs, as well as the ontology accelerators, have been provided [11]. MAS ontologies are kept in the encyclopedia base, the ontological board, and the ontology editor, as illustrated in Fig. (2). MAS ontologies convert and modify new ontologies retrieved from the ontological board, and then modify this additional taxonomy to common ontologies that may be read by all agents of MAS Ontology.

```
Ontology Structure
Thing
        CBR DESC
              UAP (University Admission Process)
              Grade of Univ cases type
        CBR INDEX
              Params
              Courses
              Grade Univ
              Fee
              Criteria
              University
              Semesters
              Faculties
        CBR_CASE
                    UAP (University Admission Process)
              Grade of Univ cases type
```

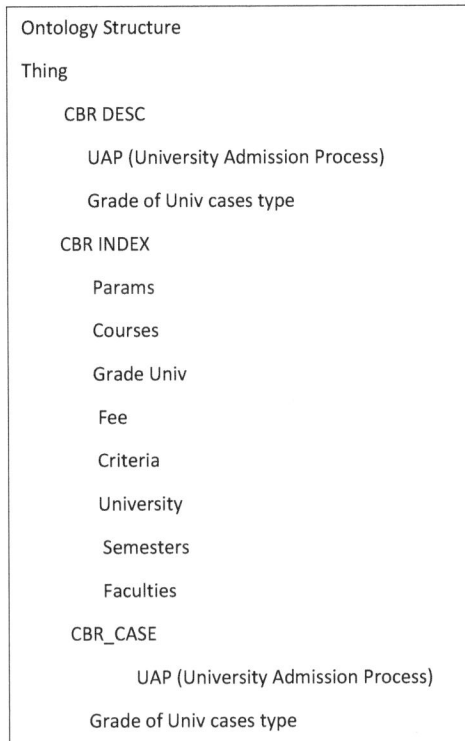

Fig. (2). Ontology-based MAS's.

AGENT ONTOLOGY INTERACTIONS

It is possible to add knowledge-level signals in interactions with ontologies and ontology enhancers. Agents' ability to adapt their communication protocols based on their current situation is a recurring issue [12]. In this section, we provide a CPN-based method for creating flexible protocols for interaction. In the first part, we quickly present the fundamental ideas of CPNs, and in the second, we show how CPNs may be used for agent engagements.

Petri Nets and Coloured PN's

Tokens are displayed in the 4-tuple depicted as a collection of Places (P1, P2, P3), Transitions (T1 to P1 and P2 to T1), and Arcs (T2 to P1). This 4-tuple may quantitatively describe the basic construction of a PN. P1 and P3 each carry a single token in the starting condition. A system's net architecture and discharge rules determine how it transitions between states [13]. Transitional firing regulations for various types of PN's are not the same. When they fire, all PNs, on the other hand, do the same thing: A transformation may be activated if the token quantity for all input locations is more than or equivalent to the strengths of their arcs. Transitions are collections of non-empty types, commonly known as colored sets; tokens in the transition's input positions will be shifted to the transition's output places when it is activated. It's a list of transformations; it's an Arcs collection; it's the node utility, the color function, the guard one, the expressions one, and the introduction one. P is the array of locations and T is the list of transitions.

Instead of being blank indicators like PNs, CPN tokens include data [14]. Tokens may be found in a variety of locations where CPNs are present. CPN arcs may specify the types of currencies that can be exchanged as well as the conditions under which the tokens can be exchanged. Multi-set component departure and arrival may be identified using an appropriate constraints function. When a CPN transition occurs, token restrictions are imposed by guard mechanisms.

Agent Protocols

According to an increasing corpus of research, CPNs may be used to more properly mimic agent interaction protocols. As a network of components carrying protocol and policy for interaction, CPNs may be used to represent an agent-interaction protocol. In order to simulate the states of an agent communication protocol, CPN locations are utilized. Every location has a predetermined set of data types that may be kept in its vaults. The colors of the symbols indicate the value of the data that is communicated between agents. The protocol's interaction rules are conveyed through the CPN movements and their corresponding arches. As long as the color limits set out by an arc are adhered to, tokens are necessary for a transition to take place [15]. To put it another way, if this transition is activated, it may have the stated consequences. A transition uses a conversation policy and new tokens are introduced to all output locations while using all input tokens as calculation variables. A protocol's state (marking) changes and the protocol enters a terminal state if there are no active or fired transitions.

To demonstrate how CPNs may be used to simulate agent interaction protocols, consider the FIPA inform protocol (FIPA, 2004). Modeling FIPA requests as CPNs is shown in Fig. (**3**). There are five various phases of interaction, and each one is shown in a different area. The phase "inform" will be activated if tokens in the "Start" location fulfill the "send" arc's constraints (belonging to the data type "message"). This transition will replace tokens from the "Start," "Received," and "Terminated1" locations. "Process inform" is triggered as soon as "Receive" receives a token. The interaction ends when the "process inform" transition is triggered [16].

ACL Message
Sender
Receiver
Reply to
Communicative act
Language
Encoding
Ontology
Protocol
Conversation-ID
In reply to:
Reply with:
Reply by:
User Properties

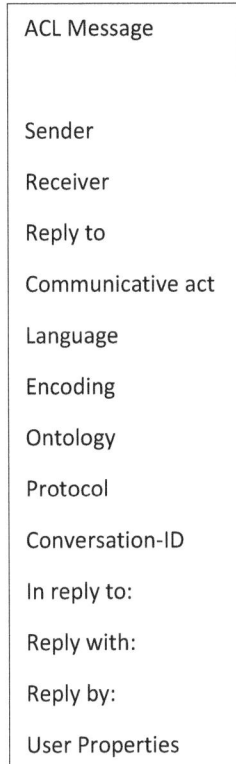

Fig. (3). FIPA inform protocol.

MULTI-AGENT INTERACTION WITH COMMUNICATION

As was said in the first part, the vast majority of agents have interaction protocols hard-coded into them. Due to the fact that pre-coded protocols are difficult to update at runtime, this trait reduces the adaptability of agent engagements. We'll teach you how to empower your agents to make new relationships with CPNs. Agents do not have to follow a set protocol in order to interact with one another. This means that agents will be able to create and modify their own interaction protocols while interacting, and they can do so in real-time. Specifying and analyzing protocols are the important ingredients of this method.

Sending and Receiving Protocol Specifications Default Protocol

A system's default method of interacting with its agents is shown in Fig. (4). Two PS locations and one transition make up the default protocol. This is how the standard interaction protocol works, according to the following explanation:

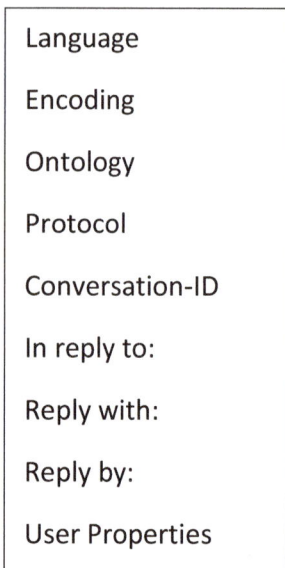

Fig. (4). Interaction protocol.

It's a form of data known as PS (protocol specification). A PS token contains a protocol description that outlines the rules of a particular interaction. In order to better understand the standard's structure, we've included Fig. (**5**).

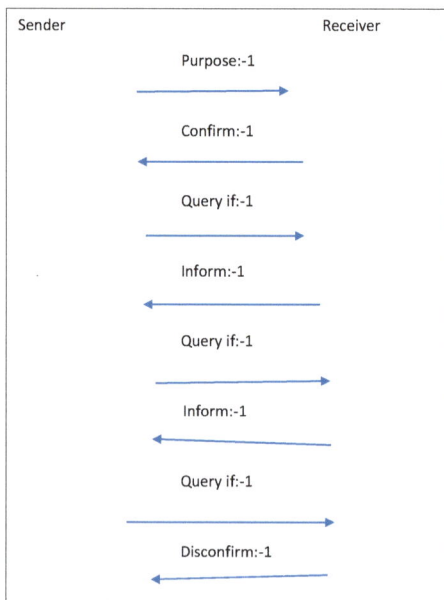

Fig. (5). Inform protocol.

PS tokens are exchanged at the CI site according to the standard protocol. A PS token in a CI may be used to establish communication between two agents. A PS-set location (protocol specification set) is a location that accepts PS tokens supplied by other agents in the default protocol and defines its preferred method of contact [17].

SPS transitions are used to pass forth protocol information while utilizing the default protocol. An SPS metamorphosis takes two input and generates two outputs: a CI position and a PS-Set position. Enabling the SPS conversions is an option if a token is available at the CI site. After the SPS changeover, tokens from the CI will be moved to the PS-Set location [18].

In the beginning, it is important to define a communication system. In order to interact with other agents, agents produce PSs and store them in their CIs for future use. PS has an interaction description standard based on the CPN paradigm. It allows the agent to indicate where the other agent must input a token to indicate what data it needs from that agency (s). There are two ways to provide data type and data value restrictions in arcs: by using constraint functions and by using the associated place type. There are no restrictions on how the requirement may be shown; in Fig. (**6**), the corresponding CPN model is shown in several presentation styles.

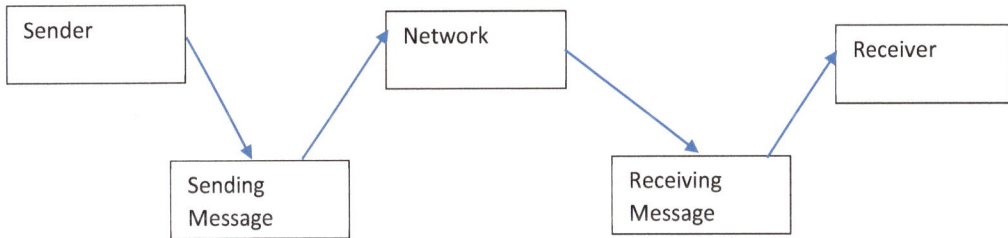

Fig. (6). CPN model of inform protocol.

Protocol Analysis

There are several factors that influence whether an agent is capable of participating in an interaction, including its current condition. The Petri Net theory may be used for PN models in several ways, one of which is to establish whether they can be run. In this chapter, the interaction protocols obtained by an agent are evaluated using matrix equation techniques. Defining a few terminologies can help us better comprehend the protocol analysis technique [19].

A place type (PT) and a place attribute (PA) are two separate sorts of places (PA).

In the second definition, when an agent acquires a PS, they gain access to a PTS (Place Type Set). UTS is a set of data types that an agent is able to interpret since they are stored in its level of knowledge. Initial network security steps include determining if an agent can comprehend what has been sent. Understandable type sets (UTS) are defined in the following section for agents with an intelligible type set (UTS) and a PS:

There are three sets of type sets: PS place type set (PTS), agent type set (UTS), and non-understandable type set (NTS). The agent can grasp the PS if all three of these factors are equivalent. During the place classification method, the agent generates an NTS based on its UTS and the PTS of the PS token. Step 2 of deep packet inspection will go forward if agents can understand PS and hence create an empty UTS. With a letter attached that describes the issue and suggests a potential solution, it will be returned to the sender. There are two PS symbols in Agent 3's PS-set. P1 and P2 are two PS tokens that were provided by agents 1 and 2, respectively. It can be seen in the following table that Agent 3 has a UTS of 1. We may deduce that P1 can be understood, but P2 cannot be understood, since DataType7 is not in the UTS of Agent 3. Thus, P2 will be sent back to Agent 2 with the following message to Agent 2: "DataType7 is un-understandable".

Interaction Analysis

For deep packet inspection, the second step is to determine whether or not the PS may be accepted by the agent or not as it would interfere with its primary objective. "State checking" is the term for this practice. Mathematical representations of PN models may be represented as matrix equations. An illustration of this shows a centralized system for the D+ and Dare values that characterize the PN model's output and input functions, respectively. For example, the nth-place token number may be denoted by a PN model's marking set (m1, m2, *etc.*). It is possible to depict the marks as follows: (1, 0, 1, 0). For situations involving interaction modeling, we provide the following formulations based on the matrix equations methodology:

In the PN model, an engagement protocol's matrix is represented by IM Matrixes. The incoming and outgoing parameters of the PN model are represented by IM- and IM+. If IM+ and IM- are added together, the result is IM.

When an agent wants to utilize a PS, he or she must provide the PS with the needed token set (RTS).

It's the total number of points an agent has earned *via* interactions with other PS agents that we're referring to when we say "gain token set."

How much GTS will be generated from an exchange may be determined by using the IM in a communication protocol. Additionally, the agent may assess whether the interaction's intended outcome is thwarted because of the viewpoint consequence. Crypto currencies of DataType1 will be lost if Protocol-1 is authorized for special data P4 and P5, but credits of DataType2 would be gained. It is up to Agent-1 to determine whether the message is in its better interests.

MULTI-AGENT INTERACTION APPLICATIONS

As a second stage in deep packet inspection, you must assess whether the PS may be accepted by the agent or not, as doing so would compromise the agent's primary goal. This process is referred to as "state checking" in the industry. Matrix formulas may be used to describe computational models of PN models. A centralized system for the D+ and Dare values, which describe the output and input components of the PN model, is an instance of this. For example, the labeling set of a PN model may serve as a symbol for the nth-place token number (m1, m2, *etc.*). The markings may be shown in this manner: It is (1, 0, 1). Formulations based on the matrices equations technique are provided for scenarios in which interactions modeling is required:

The IM matrix represents an engagement procedure in the PN model. The input and output parameters of the PN model are represented by the IM- and IM+ matrixes. If IM+ and IM- are added together, the result is IM.

The necessary token set must be provided to the PS by the agent before the PS may be used (RTS).

When we say "gain token set," we're talking about the total amount of points an agent has earned *via* interactions with other PS agents.

We use the IM in a communication protocol to calculate how much GTS an exchange will create. Another consideration is if a potential outcome of an encounter conflicts with its own objectives. A token of DataType1 will be lost and two tokens of DataType2 will be gained if Protocol-1 is accepted for data types P4 and P5. Agent-1 will weigh the pros and cons of participating in this communication before making a final decision.

CONCLUSION

In a multi-agent system, an agent's social skills are put to the test. Protocols for MAS agent contact, particularly in open contexts, limit the adaptability of the agent-to-agent interaction. We have outlined a strategy for letting agents create knowledge-level interaction protocols in this chapter. If the received protocol is not comprehensible, or if it disagrees with an agent's aims, this strategy may also be used. Using these characteristics, agents are able to pick or develop appropriate protocols for open working environments.

REFERENCES

[1]   Q. Bai, and M. Zhang, "An ontology-based multi-agent auction system", In: *Proceedings of the International Conference on Intelligent Technologies,* Chiang Mai, Thailand, 2003, pp. 197-202.

[2]   K.S. Kaswan, and J.S. Dhatterwal, "Chapt. 1: The Use of Machine Learning for Sustainable and Resilient Buildings", In: *Digital Cities Roadmap* John Wiley & Sons., 2021, pp. 1-62.
[http://dx.doi.org/10.1002/9781119792079.ch1]

[3]   Q. Bai, and M. Zhang, "SWOAM — A semantic Web-based online auction multi-agent system", In: *Proceedings of the International Conference on Intelligent Agents, Web Technologies, and Internet Commerce,* Gold Coast, Australia, 2004, pp. 329-336.

[4]   R. Cost, "Modelling agent conversations with coloured petri nets", In: *Working Notes of the Workshop on Specifying and Implementing Conversation Policies* Seattle WA, 1999, pp. 59-66.

[5]   S. Cranefield, M. Purvis, M. Nowostawski, and P. Hwang, "Ontology for interaction protocols", *Proceedings of the 2nd International Workshop on Ontologies in Agent Systems — AAMAS'02,* 2002 Bologna, Italyhttp://sunsite.informatik.rwth-aachen.de/Publications/CEUR-WS//Vol-66/oas0216.pdf

[6]   S Srinivasan, and Kumar Vivek, "Multi-agent-based decision Support System using Data Mining and Case Based Reasoning", *International Journal of Computer Science Issues.,* vol. 8, no. 4, 2011.

[7]   T. Finin, Y. Labrou, and J. Mayfield, "KQML as an agent communication language", In: *Software agents.,* J.M. Bradshaw, Ed., MIT Press: Cambridge, MA, 1997, pp. 291-316.

[8]   S.P. Yadav, and S. Yadav, "Fusion of Medical Images in Wavelet Domain: A Discrete Mathematical Model", In: *Ingeniería Solidaria* vol. 14. Universidad Cooperativa de Colombia: UCC., 2018, no. 25, pp. 1-11.
[http://dx.doi.org/10.16925/.v14i0.2236]

[9]   V. Horro Vashisht, A. K. Pandey, and S. P. Yadav, "Speech Recognition using Machine Learning", In: *IEIE Transactions on Smart Processing & Computing* vol. 10. The Institute of Electronics Engineers: Korea, 2021, no. 3, pp. 233-239.
[http://dx.doi.org/10.5573/IEIESPC.2021.10.3.233]

[10] Kaswan Kuldeep Singh, Singh Dhatterwal Jagjit, and Balyan Anupam, "Intelligent Agents Based Integration of Machine Learning and Case Base Reasoning System", In: *IEEE Conference "2nd International Conference on Advance Computing and Innovative Technologies in Engineering* Galgotias University: Greater Noida, 2022.
[http://dx.doi.org/10.1109/ICACITE53722.2022.9823890]

[11] K. Jensen, An introduction to the practical use of colored petri nets. *Lectures on Petri Nets II: Applications. Lecture Notes in Computer Science* vol. 1492. Springer-Verlag: Berlin, Germany, 1998, p. 237292.

[12] K.S. Kaswan, J.S. Dhatterwal, and K. Kumar, "Blockchain of internet of things-based earthquake alarming system in smart cities", In: *Integration and Implementation of the Internet of Things Through Cloud Computing.* IGI Global, 2021, pp. 272-287.
[http://dx.doi.org/10.4018/978-1-7998-6981-8.ch014]

[13] V.R. Lesser, "Cooperative multiagent systems: a personal view of the state of the art", *IEEE Trans. Knowl. Data Eng.,* vol. 11, no. 1, pp. 133-142, 1999.
[http://dx.doi.org/10.1109/69.755622]

[14]   M. Nowostawski, M. Purvis, and S. Cranefield, "A layered approach for modelling agent conversations", In: *Proceedings of the 2ⁿᵈ International Workshop on Infrastructure for Agents, MAS, and Scalable MAS*Montreal, Canada, 2001, p. 163170.

[15]   A. Baliyan, J.S. Dhatterwal, K.S. Kaswan, and V. Jain, "Role of AI and IoT Techniques in Autonomous Transport Vehicles", *Transactions on Computer Systems and Networks,* vol. 5, pp. 1-23, 2022.
[http://dx.doi.org/10.1007/978-981-19-2184-1_1]

[16]   D. Poutakidis, L. Padgham, and M. Winikoff, "Debugging multi-agent systems using design artefacts: The case of interaction protocols", In: *Proceedings of the 1ˢᵗ International Joint Conference on Autonomous Agents and Multi Agent Systems*Bologna, Italy, 2002, pp. 960-967.

[17]   A.S. Rao, and M.P. Georgeff, "An abstract architecture for rational agents", In: *Proceedings of the 3ʳᵈ International Conference on Principles of Knowledge Representation and Reasoning*San Mateo, CA, 1992, pp. 439-449.

[18]   J.S. Dhatterwal, S. Dixit, and S. Srinivasan, "Implementation of case base reasoning system using multi-agent system technology for a buyer and seller negotiation system", *Int J Electron Commun Engg,* vol. 7, no. 3, pp. 63-67, 2021.

[19]   S. P. Yadav, and S. Yadav, "Mathematical implementation of fusion of medical images in continuous wavelet domain", *Journal of Advanced Research in dynamical and control system,* vol. 10, no. 10, pp. 45-54, 2019.

CHAPTER 2

Strengthening Corporate Social Responsibility Practices through Artificial Intelligence

A. Menaga[1], Yasmeen Bano[2], Narendranath Uppala[3] and S.Vasantha [1,*]

[1] *School of Management Studies, Vels Institute of Science, Technology & Advanced Studies (VISTAS), Chennai, India*

[2] *School of Management Studies ,Sathyabama Institute of Science and Technology (SIST), Chennai, India*

[3] *Putra Intelek International College, Petaling Jaya, Malaysia*

Abstract: Artificial intelligence (AI) has gained enormous usage in business in recent years. Still, in regard to measuring business ethics and morality, otherwise called corporate social responsibility, the use of Artificial Intelligence is limited to a greater extent. In this regard, the purpose of the study is to conceptually formulate the implementation of AI in CSR programs. For gathering data, the study utilised a structured questionnaire. Employers from a range of governmental and commercial organisations provided the primary data for the study. Using AMOS 21's Structural Equation Modelling (SEM) and SPSS 21, the projected model was empirically tested. The Research concludes that AI can strengthen effective CSR practices. The research also uses SEM to establish a cause-and-effect connection between the research variables.

Keywords: Artificial Intelligence, Behaviour Analysis, Corporate Social Responsibility, Fraud Detection.

INTRODUCTION

To delve into the conceptualization of efficient corporate social responsibility, it is imperative to grasp the fundamental essence of Artificial Intelligence (AI). AI entails the utilization of computer systems to replicate human cognitive processes and make decisions or solve problems based on trained algorithms. It should be noted that AI is limited to mimicking human intelligence and thus falls short in the domain of corporate social responsibility, as humans inherently possess greater levels of responsibility compared to machines. However, AI can play a

---

* **Corresponding author S.Vasantha:** School of Management Studies, Vels Institute of Science, Technology & Advanced Studies(VISTAS), Chennai, India; E-mail: vasantha.sms@velsuniv.ac.in

Adarsh Garg, Valentina Emilia Balas, Rudra Pratap Ojha & Pramod Kumar Srivastava (Eds.)

significant role in assessing and mitigating corporate social irresponsibility, thereby enhancing the effectiveness and efficiency of corporate social responsibility initiatives. There is only a limited literature review on why machines cannot be responsible as much as human beings [1].

The organisation is a part of business for social responsibility; acting responsibly is one of the drivers for firms to showcase themselves as reputed organisations. To prove this, the firm is working on a sustainable CSR program that makes an organisation portray good. However, this comes with some underlying critics. Corporates generally focus on showcasing the better side and the critical aspects. Critics are a vital part that reduces the effectiveness of CSR. To address this criticism, automation is a way to improve the CSR process.

Critical criticism of CSR:

- Behavioural issues: Intention of the manager or management to ignore CSR but to enjoy the economic value the firm has gained.
- Fraudulent: Lying and cheating on Organisation returns or Greenwashing products.
- To address this issue, the research has framed the following objectives for Effective CSR performance.

OBJECTIVE OF THE STUDY

- To review past literature based on the relationship between corporate social responsibility and Artificial Intelligence.
- Develop a conceptual framework for the AI-based CSR model.
- To evaluate integrated AI toward efficient CSR practice.

LITERATURE REVIEW

Applying Artificial Intelligence to Corporate Social Irresponsibility

The crucial idea is that AI cannot perform corporate social responsibility. Instead, AI can help reduce unethical or immoral activities, otherwise called corporate social irresponsibility. According to a study [2], the primary placement of integrating AI into CSR is to train AI for the following measures:

- Measurement of Materiality
- Measurement of Performance

To understand the basic concept of materiality, an organisation that performs CSR will confront economic, social, and environmental issues that apply to sustainable business. A methodical materiality evaluation determines which subject matter should be considered in an industry or sustainability strategy [3] AI can help to assess the stakeholder, meaning AI can help in determining the most powerful or influential stakeholder who can add value to the business. According to a study [4], a mathematical model that is used to resolve MULTI-CRITERIA DECISION-MAKING (MCDM) problems can be utilised to determine preferences and perform the materiality evaluation [5]. The development of materiality assessment has also included failure modes and impact analysis [6]. Measurement of Organisation ESG performance by human power is quite complex; that is, whether AI can be used to measure the firm ESG or CSR performance; AI can use data to track the CSR performance and use processing and quality assurance, then it will adopt the analytical model like sector-based indication weightage and finally through output, provide the Ranking, as shown in Fig. (**1**).

When incorporating the Corporate Social Responsibility (CSR) and Artificial Intelligence (AI) into business operations, it is crucial to be aware of the potential pitfalls to be avoided. To ensure a successful implementation of this integrated approach, it is essential to steer clear of the following errors. There are two categories in which AI can be segregated, *i.e.* Strong AI and Weak AI [7], so when it comes to measurement and suppression of corporate social irresponsibility, AI can perform much more strongly than humans. Regarding actual CSR initiatives, Artificial Intelligence (AI) falls short in its ability to outperform humans in moral activities, resulting in relatively weaker performance in such tasks. Consequently, future research endeavours should focus on addressing the genuine ethical capabilities of AI, seeking to enhance its ethical performance.

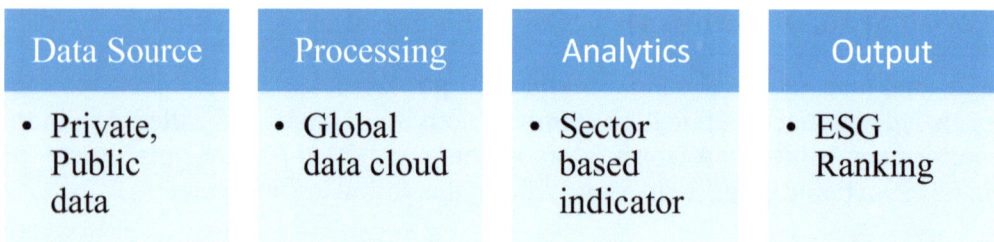

Data Source	Processing	Analytics	Output
• Private, Public data	• Global data cloud	• Sector based indicator	• ESG Ranking

Fig. (1). Performance Measurement model [6].

Traditional Corporate Social Responsibility

Conventional corporate social responsibility is all about human performance. Undoubtedly many firms are performing well. The proposed research paper aims to address the gaps emphasized by critics of Corporate Social Responsibility (CSR). It acknowledges that there are situations where individuals, who are integral parts of corporate entities, may not bear sole responsibility for societal and environmental concerns. Irresponsible behaviour on part of the humans across different dimensions can be attributed to personal mood factors and predilections [1].

The subsequent points outline various manifestations of corporate imprudence: There are situations where it is not feasible for corporations to bear full responsibility for certain outcomes or circumstances.

- Some corporations may reject the assertion as mentioned above, considering it unacceptable or unfounded based on their specific circumstances or perspectives. Impossibility of being responsible.
- Unacceptable to the above claim.

The phenomenon, Impossible to be responsible, happens (i) when there is no clarity among those responsible. (ii) Responsible act is considered a doubtful occurrence, *i.e.*, one single person from the organisation cannot be blamed for being irresponsible because corporate social responsibility is not an individual free will to choose a course of action [8]. According to [9], he defines humans are prone to "The temptation of innocence, which means every human will try and avoid the consequences of their actions. He states the temptation of innocence into two categories (i) victimisation and (ii) Infantilism. Both express the mistakes that happened because of carelessness and the person should be free from guilt.

Core activities of an organization are predefined. Every individual has a set of standards to perform their job. Corporate social responsibility is one such predefined action that an individual must complete the tasks that the management sets and in the location of activities with standard procedures, so individual freedom is not so encouraged [10]. A good pre-defined action must contain the following things:

- The value of stakeholders must be considered.
- The needs and wants of the end customer should be analysed and catered to accordingly.
- Businesses should perform ethically even if it is not an obligation.

Based on the above, the following aspects can be fulfilled by a professional; a professional is a person who is technically trained in the required field. These people can bring a lot of technological advancements and creativity to the organisation [11]. On the other hand, a professional must understand the core aim of the business and should perform the job efficiently. The following are the set of unsaid standard professional procedures, but it is still not blameable on the professional.

Occurrences of mistakes by a professional while performing business activities are not considered illegal because it is claimed to be an innocent occurrence. This is also called corporate social irresponsibility (CSI); AI can be implemented into the business to reduce this CSI.

A vast majority of organisations practice part CSR and CSI. CSR is an activity performed for the welfare of the stakeholder and the environment. In contrast, CSI is the opposite of the performance of activities in the interest of the owner and the organisation. CSR and CSI can be performed at the same time by one firm, *e.g.*, Coca Cola water depletion CSI issue in India, whereas at the same it is serving the greatest in Croatia. Worst of all, CSI is doing illegal business. CSR report is an extensive fabrication of hidden CSI in the organisation [12].

Corporate social irresponsibility cannot be eliminated from the business, say it is in the public or private sector. As long as there is human existence, little research supports CSI to a certain percentage that a small amount of CSI is good for society.

There are limited ways in which CSI can be reduced by the firms, the widespread method is the detection of fraudulence and analysis of such behaviour, if this is not implemented in future years, CSR practice will become a complete fabrication, and AI can be integrated to reduce CSI.

Fraud Detection

Fraud detection will come under intelligence and cognitive CSR; machine learning acts as a powerful tool to eradicate troublesome activity and brings up a revolution in intelligence and Cognitive CSR helps to improve the CSR process in a better manner. The intelligence and cognitive mechanism understand the core value of the business and are able to bring out a positive output for its internal and external members [13]. AI suggests the Program Strategy for better organizational improvement. On the other hand, with regard to CSR, it indicates the organisation has an optimized program from the available list.

Artificial Intelligence (AI) not only enables the assessment of organizational performance but also possesses the capability to identify recurring activities that are contrary to the management's vision or violate regulations and global standards. For instance, AI can detect instances such as environmental pollution or self-serving practices in product manufacturing, such as water depletion by companies as in the case of Coca-Cola. When such activities are detected, AI acts as a red flag indicator for the organization, allowing investors and regulatory bodies to take appropriate actions to address these issues. Artificial intelligence can create fear and behavioural control for the organisation from going off track from illegal or standard activities [14]. So other activities the AI can detect are misalignment incentives among shareholders and managers or organisations trying to retain. According to [15] if a firm fails to align and integrate, the organisation's strategy and CSR program will create a lack of competitive advantage. Fraud detection can be done with automated data entry clerical processes. This will give quick suggestions to change the action. If not, it will prompt an error in which the organisation's ESG ranking will go down automatically.

Behavioural Analysis

To a great extent, CSR is miscarried because of managerial behaviour.

Following are some managerial behaviour issues that will hinder CSR success [16].

- Incentive misalignment: When there is a disconnect between incentives, managers may withhold funds and discontinue Corporate Social Responsibility (CSR) initiatives.
- Conflicts with CSR agency: Conflicting interests or misalignment between an organization and its designated CSR agency can hinder the successful implementation of CSR activities.
- Selfish behaviour and profit-driven temptation: Instances of self-centered conduct or the allure of maximizing personal profit can lead to a disregard for shareholders and their interests, prioritizing personal gain over responsible business practices.

CSR efforts have drawn criticism for being dishonest, self-centered, and commercial [17].

The following are some of the techniques AI can improve on behavioural issues.

- Information extraction algorithms to automatically extract crucial data from reports on the effectiveness of behaviour change interventions; and
- Algorithms for machine learning and reasoning infer the extracted data to produce proposed interventions for anomalous circumstances and forecast the expected results of interventions that have not yet been tried.

While AI cannot fundamentally alter human behaviour, it does possess the capability to influence behaviour to a limited extent. One effective technique for behaviour change is the implementation of positive reinforcement. For instance, if a manager or employee engages in a CSR action, they can be acknowledged and rewarded with additional incentives or elevated positions within the organization. Natural Language Processing (NLP) plays a crucial role in gathering and analysing information regarding employees' past and current performances in CSR activities. This enables the identification of improvements, resulting in improved rankings, while a lack of action is reflected as "NIL" in the system.

METHODOLOGY

The study adopted a structured questionnaire for evaluating the effect of Artificial Intelligence on efficient CSR practices as depicted in Fig. (**2**). There are 24 items in the study which are measured through a 5 points Likert scale. The study has used the construct "Integrated AI" as a mediating variable. Convenience sampling techniques were used in this study and descriptive research was followed.

PROPOSED CONCEPTUAL MODEL

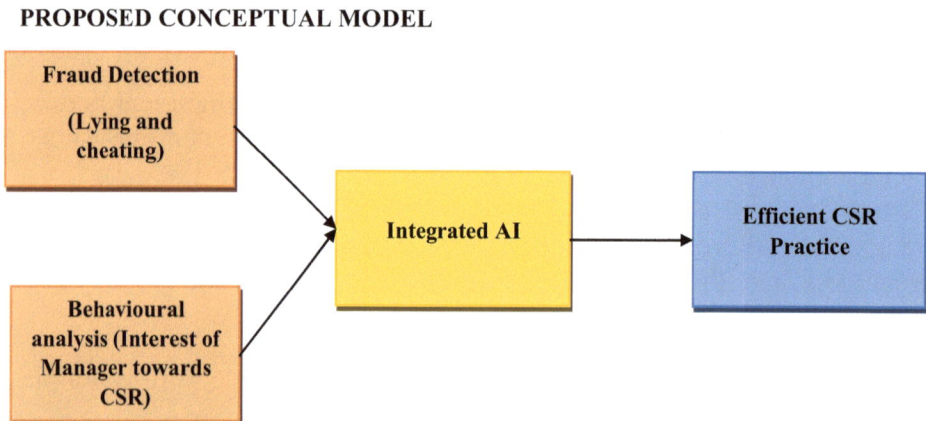

Fig. (2). Proposed Conceptual framework for strengthening CSR practices through AI.

After collecting the data from the employers of public and private organizations, different types of numerical approaches were applied to test the hypothesis of this study, which includes data screening, checking whether all data has been entered correctly, identifying missing data, outlier checking, and normality.

The study's data sources included 270 employers from different publicly traded and private businesses. The data were examined using SPSS, statistical software for social sciences, and AMOS. The validity of the questionnaire was evaluated for ascertaining its reliability, and the outcome was found to be 0.879. To prove the existence of a relationship between the variables under consideration, structural equation modelling (SEM) is used (Fig. **3**).

DATA ANALYSIS AND RESULTS

H_1: Fraud detection is Positively associated with Integrated AI.

H_2: Behavioural analysis is Positively associated with Integrated AI.

H_3: Integrated AI is positively associated with Efficient CSR practices.

H_4: There is a causal relationship among the study variables (Fraud detection, Behavioural analysis, Integrated AI, Efficient CSR practices).

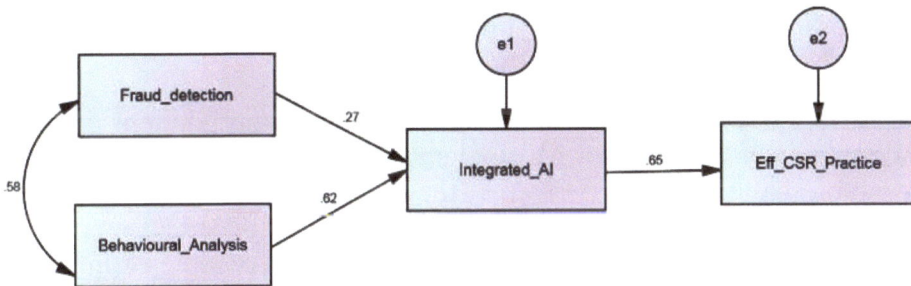

Fig. (3). SEM Model for strengthening CSR practices through AI.

I. **Observed, endogenous variables:** Fraud detection, Behavioural analysis, Integrated AI, and Efficient CSR practices.
II. **Unobserved exogenous variables:** e1 and e2

Using a PATH ANALYSIS MODEL that was developed and assessed using AMOS, the relationship between the endogenous components and the exogenous variables was investigated. The suggested model shows how path analysis may be used to determine whether the model overall fits the data and to determine whether the variables are significantly correlated.

Table **1** clarifies the causal relationship among the AI factors (Fraud detection, Behavioural analysis, Integrated AI with Efficient CSR practices.

Table 1. Variables used in Structural Equation Model.

Dependent Variables		Independent Variables	Unstandardized Coefficient	S.E.	Standardized Coefficient	T Value	P - Value	Result of Hypothesis
Integrate AI	<---	Fraud_detection	0.288	0.048	0.269	6.05	0.000	H1 is Supported
Integrated_AI	<---	Behavioural Analysis	0.618	0.044	0.618	13.88	0.000	H2 is Supported
Eff_CSR_Practice	<---	Integrated_AI	0.611	0.043	0.653	14.14	0.000	H3 is Supported

The independent variable fraud detection shows the relationship with the dependent variable integrated AI. Integrated Fraud detection has a positive influence on AI, with an unstandardized coefficient value of 0.288. According to the anticipated positive indication, the effect is favourable and for every increase in Integrated AI unit, fraud detection will increase by 0.288 times. For this study, a standard estimate of less than 5% is suitable. T-values must be either more than +2 or less than -2. The variable's t-value is 6.05, which is higher than +2. The variable will have more faith in the coefficient as a predictor, the higher the t-value. Therefore, the hypothesis "Fraud detection is positively associated with Integrated AI" is accepted.

The independent variable behavioural analysis shows the relationship with the dependent variable integrated AI. Integrated AI is positively influenced by behavioural analysis, with 0.618 as the unstandardized coefficient value. According to the expected positive sign, which indicates the effect is positive, the behavioural analysis will increase by 0.618 times for every unit rise in Integrated AI. The unstandardized coefficient value was found to be significant at 1% level of significance since the p value was less than 0. 05. T-values must be either more than +2 or less than -2. The variable's t-value is 13.88, which is higher than +2. The variable will have more faith in the coefficient as a predictor of the higher t-value. As a result, it is accepted that "Behavioural analysis is positively connected with Integrated AI".

The Integrated AI predictor shows the relationship with the dependent variable of Efficient CSR practice. Mediating factor of Integrated AI has a positive effect on Efficient CSR practice with an unstandardized coefficient value of 0.611. The projected positive sign suggests that this effect is favourable, for every unit rise in Efficient CSR practise, the Integrated AI will increase by 0.611 times. The unstandardized coefficient value was determined to be significant at 1% level of significance since the p value was less than 0.05, *t*-value greater than +2 or less than – 2 is acceptable. t-value for the variable is 14.14, which is greater than +2. The higher the t-value, the greater the confidence the variable will have in the coefficient as a predictor. Therefore, the hypothesis "Integrated AI is positively associated with Efficient CSR practice" is accepted and confirms that there is a partial mediating role of Integrated AI between the independent variables (Fraud detection, behavioural analysis) and dependent variable of Efficient CSR practices in this study.

According to Table **2**, the alternative hypothesis (H4) is accepted because the computed P value is 0.000, which is less than 0.05, and the model has a decent fit. The GFI and AGFI values in this case are both more than 0.9, indicating a strong fit. A perfect fit is predicted by the anticipated CFI value of 0.916, and the fit's accuracy was further supported by the RMSEA value, which was found to be 0.056 and less than 0.08. A CMIN/Df of fewer than 5 is acceptable, according to [18]. The proposed model is legitimate, according to the value reported in the table. Additionally, the proposed model's exact match with the data set is guaranteed.

Table 2. Model Fit Summary for Structural Equation Model.

Goodness of Fit Statistics	Value	Values for Good Fit
P -VALUE	0.000	<0.05 (Hair *et al*.,1998)
CHi SQUARE / DF	3.659	<5.00 (Hair *et al*.,1998)
GOODNESS OF FIT INDEX	0.920	>0.90 (Hu and Bentler,1999)
ROOT MEAN SQUARE ERROR OF APPROXIMATION	0.056	<0.08 (Hair *et al*.,2006)
ADJUSTED GOOD OF FIT INDEX	0.976	>0.90 (Hair *et al*. 1998)
COMPARATIVE FIT INDEX	0.916	>0.90 (Hu and Bentler,1999)
NORMED FIT INDEX	0.913	>0.90 ((Hu and Bentler,1999)

FINDINGS AND DISCUSSION

In the modern world, artificial intelligence (AI) has made a significant impact on all facts of business and human activity. By utilizing data produced from a massive increase in digital touchpoints, it holds the promise of fostering efficiency and effectiveness.

This study has proved that fraud detection is positively associated with Integrated AI. Hence, H1 of this study is accepted [19]. A study discusses AI, arguing that, with the use of robots, the process of giving caregivers greater flexibility in their everyday lives can be enhanced. They base this claim on a study of the competencies approach.

H2 has been accepted because the direct relationship between behavioural analysis and effective Integrated AI is significant in the study. According to a study [20], computer systems can play a significant moderating role in human agency, strengthening moral autonomy.

H3 is also accepted and indicated that the direct relationship between Integrated AI and Efficient CSR practices is significant in the study. Corporate social responsibility (CSR), which is employed in a wide range of domains such as areas of policy, programs, and action when interacting with stakeholders, has become the key concern of firms in the worldwide marketplace. From a CSR aspect, businesses must embrace the idea of being socially responsible while integrating AI into daily operations [21].

CONCLUSION

Artificial intelligence (AI) is an extremely featured tool in modern science—a tool with nearly limitless merits and demerits. Currently, the human perspective and the AI perspective can be used to roughly divide the ethical and social implications of AI into two separate fields of study. This distinction is made due to our imperfect grasp of its function rather than the AI's primitive character as a potentially detached (in logic and behaviour) entity. Furthermore, fast computation is a widely known benefit of Artificial Intelligence and machine learning. It develops an understanding of a user's app usage patterns like transaction methods, payments, *etc.*, which can easily detect an anomaly in real-time. With better efficiency than manual methods, it avoids the occurrence of false positives and allows specialists to focus on more complex issues.

REFERENCES

[1]   B. Krkač, and I. Kristijan, Artificial intelligence and social responsibility. *The Palgrave Handbook of Corporate Social Responsibility,* David Crowther, Shahla Seifi, Eds., Springer, 2020, pp. 1153-1175.

[2]   G. Li, N. Li, and S.P. Sethi, "Does CSR reduce idiosyncratic risk? Roles of operational efficiency and AI innovation", *Prod. Oper. Manag.,* vol. 30, no. 7, pp. 2027-2045, 2021.
       [http://dx.doi.org/10.1111/poms.13483]

[3]   Yadiv S. P., D. P. Mahato, and N. T. D. Linh, *Distributed Artificial Intelligence.* CRC Press., 2020.

[4]   S. Edelkamp, and S. Schrodl, *Heuristic search: theory and applications.* Elsevier, 2011.

[5]   P. Calabrese, and J. Cardy, "Quantum quench in 1+ 1 dimensional conformal field theories", *J. Stat. Mech.,* vol. 6, p. 064003, 2013.

[6]   V. Bellomo, Z. W. LIM, G. Nguyen, Q. D. XIE, M. Y. YIP, and D. S. TING, "Artificial intelligence using deep learning to screen for referable and vision-threatening diabetic retinopathy in Africa", *Lancet Digit. Health,* vol. 1, no. 1, pp. e35-e44, 2019.
       [http://dx.doi.org/10.1016/S2589-7500(19)30004-4]

[7]   J. N. NILS, *The quest for artificial intelligence.* Cambridge University Press, 2009.
[http://dx.doi.org/10.1017/CBO9780511819346]

[8]   R.J. Wallace, *Responsibility and Moral Sentiment* Harvard University Press, 1998.

[9]   S.P. Yadav, and S. Yadav, "Fusion of medical images using a wavelet methodology: A Survey", In: *IEIE Transactions on Smart Processing & Computing.* vol. 8. The Institute of Electronics Engineers of Korea., 2019, no. 4, pp. 265-271.
[http://dx.doi.org/10.5573/IEIESPC.2019.8.4.265]

[10]   V. Vashisht, A.K. Pandey, and S.P. Yadav, "Speech recognition using machine learning", In: *IEIE Transactions on Smart Processing & Computing* vol. 10. The Institute of Electronics Engineers of Korea., 2021, no. 3, pp. 233-239.
[http://dx.doi.org/10.5573/IEIESPC.2021.10.3.233]

[11]   S. P. Yadav, and S. Yadav, "Mathematical implementation of fusion of medical images in continuous wavelet domain", *J. of Adv. Res. in dynamic. and contr.sys.,* vol. 10, no. 10, pp. 45-54, 2019.

[12]   P. HAWKEN, *McDonald's report: more corporate social irresponsibility.,* 2001.

[13]   F. Vanclay, "The triple bottom line and impact assessment: How do TBL, EIA, SIA, SEA, and EMS relate to each other?", *Tools. Techn. and Approa. for Sustain.,* pp. 101-124, 2009.
[http://dx.doi.org/10.1142/9789814289696_0006]

[14]   C. Edgley, M.J. Jones, and J. Atkins, "The adoption of the materiality concept in social and environmental reporting assurance: A field study approach", *Br. Account. Rev.,* vol. 47, no. 1, pp. 1-18, 2015.
[http://dx.doi.org/10.1016/j.bar.2014.11.001]

[15]   M.E. Porter, and M.R. Kramer, "Strategy and society: The link between competitive advantage and corporate social responsibility", *Harv. Bus. Rev.,* vol. 84, no. 12, pp. 78-92, 163, 2006.
[PMID: 17183795]

[16]   C. Cennamo, P. Berrone, and L.R. Gomez-Mejia, "Does stakeholder management have a dark side?", *J. Bus. Ethics,* vol. 89, no. 4, pp. 491-507, 2009.
[http://dx.doi.org/10.1007/s10551-008-0012-x]

[17]   W. Visser, *The age of responsibility: CSR 2.0 and the new DNA of business.,* vol. 5, no. 3, 2011.*J.Busin. Sys. Gov. and Ethics.,* vol. 5, no. 3, 2011.
[http://dx.doi.org/10.15209/jbsge.v5i3.185]

[18]   J.F. Hair, W.C. Black, B.J. Babin, R.E. Anderson, and R.L. Tatham, Multivariate data analysis 6th Edition. Pearson Prentice Hall. New Jersey. humans: Critique and reformulation. *J. Abnormal. Psychology.* vol. 87. Pearson Prentice Hall: New Jersey, 2006, pp. 49-74.

[19]   J. Borenstein, and Y. Pearson, "Robot caregivers: Harbingers of expanded freedom for all?", *Ethics Inf. Technol.,* vol. 12, no. 3, pp. 277-288, 2010.
[http://dx.doi.org/10.1007/s10676-010-9236-4]

[20]   J. De Mul, *Cyberspace Odyssey: Towards a virtual ontology and anthropology.* Cambridge Scholars Publishing, 2010.

[21]   M. J. H. Talukder, *Conceptualising And Validating Measurement Scales For Supplier Social Responsibility* University of Canbera: Canberra Business School, 2013.

CHAPTER 3

Role of Artificial Intelligence in Healthcare Management

Amit Bhaskar[1], Pankaj Yadav[1], Savendra Pratap Singh[1,*], Vijay Kumar[1], Sambhrant Srivastava[1], Saurabh Kumar Singh[1], Brihaspati Singh[1] and Akriti Dutt[2]

[1] *Rajkiya Engineering College, Azamgarh, Deogaon, Azamgarh, Uttar Pradesh, India*

[2] *Agriculture Department, Government of Uttar Pradesh, Uttar Pradesh, India*

Abstract: Artificial intelligence (AI) has recently become one of the most heavily debated themes in the technological world. AI is active in numerous fields and now it has lately entered the healthcare sector. In addition to biomarkers, the use of AI is increasing in a variety of applications such as genetic editing, disease prediction and diagnostics, drug development, personalized treatment, and so on. Accuracy in disease diagnostics is essential for effective and efficient treatment as well as patient safety. Artificial intelligence is a wide and varied field of data, analytics and continuously evolving insights that meet the needs of the healthcare sector as well as patients. The purpose of the many subsections in this book chapter is to shed light on how AI integrated with machine learning (ML) & Deep-learning (DL) techniques operate in various disease diagnosis domains, medication discovery, medical visualization, digital health records, and electro-medical equipment.

Keywords: Artificial Intelligence (AI), Healthcare, Machine learning (ML).

ARTIFICIAL INTELLIGENCE (AI): A NEW ERA

Artificial Intelligence (AI) is a computing technology based on algorithms and used to programme self-learning from data and make precise predictions with real-time decisions using artificial neural networks (ANN), ML, Big data, data mining, robotic process automation and so on. Almost every aspect of modern life is getting influenced by the use of Big data and ML including entertainment, healthcare, and commerce. For prediction modeling, methods from ML and AI have seen a rapid rise in popularity.

---

* **Corresponding author Savendra Pratap Singh:** Rajkiya Engineering College, Azamgarh, Deogaon, Azamgarh, Uttar Pradesh, India; E-mail: savendrasingh123@gmail.com

A comprehensive quality and applicability assessment is needed to ensure the performance, safety, and usefulness of sophisticated data-driven prediction models before they are employed and widely used, despite the potential benefits of ML and AI in healthcare.

AI applications have the potential to significantly advance healthcare across the board, from diagnosis to therapy and treatment. Numerous examples of tasks where AI algorithms are outperforming humans include the analysis of medical visualization, medical images, and the similarities between symptoms and biomarkers from electronic medical records (EMRs) with the diagnosis, prognosis, and aid for treatment of the disease [1].

Artificial intelligence has the potential to greatly enhance patient care while also reducing healthcare costs. With a growing population comes a greater demand for medical care. The healthcare industry requires creative ideas to figure out how to be more effective and efficient without spending too much money. Technology can provide the answers in this situation. Rapid improvements in the state of technology, particularly in the disciplines of AI, robotics, and mechatronics, enable the healthcare industry to grow [2]. They can now accomplish human tasks more quickly, easily, and affordably thanks to AI.

Early detection and diagnosis are both heavily pivotal to AI. DeepMind Health Technology from Google combines the use of machine learning with a neuroscience system to model and design AI-enabled human brains. It aids healthcare professionals in making correct diagnosis other important decisions. Potential future applications in the healthcare system are a major driving force behind the development of AI-based technologies. There will be an annual savings of $150 billion in US healthcare costs due to the usage of AI applications by 2026. These savings can be attributed in large part to the transition from a reactive to a proactive healthcare approach, which places more emphasis on health management and less on disease treatment. This is anticipated to lead to a decrease in hospital stays, medical visits, and treatments. The market for AI and robotics-related healthcare systems is anticipated to expand quickly and may reach up to USD 6.6 billion by the year 2021 representing an increased annually compounded growth rate of 40% [3].

TECHNOLOGICAL BREAKTHROUGHS

In the past decade, there have been various technological improvements in the fields of data science and AI. Despite decades of ongoing research in AI for a variety of applications, the current scenario of AI hype differed from earlier ones.

According to Straits research from GLOBE NEWS WIRE (July 2022), the patient portal market expects to reach $11.74 billion by 2023 from past $ 2.41 billion in 2021 at a globally compounded growth rate per year of 19.23%. The demand for electronic health records (EHRs) is rising, and more healthcare payers are adopting a patient-centric strategy, which is growing the market for patient portals. The WHO reports that more than half of the world's upper-middle and high-income countries have national EHR systems in place. Governments throughout the world are focusing on creating standards, legislation, and facilities for preserving medical records, which is driving up demand for patient portals. Cloud-based services, Big data analytics and AI are most frequently used in healthcare industries, together with different patient portals for the fast diagnosis and treatments of various diseases. The sudden global spread of COVID-19 has raised the requirement for effective healthcare assistance solutions even more [4].

APPLICATIONS OF AI IN HEALTHCARE

A variety of tasks such as clinical record keeping, administrative tasks, patient engagement and other professional assistance in areas such as medical device automation, image analysis and patient monitoring can be performed by AI for medical professionals.

Numerous hospitals use AI-enabled processes as decision-support systems for medical personnel in the diagnosis and treatment of various diseases. AI systems also put an impact on organizational aspects of care delivery like improving the performance of various workflows, such as managerial and nursing activities in hospitals [5].

In 2018, Forbes & Accenture stated that the most crucial areas are connected like machines, dose error reduction, cybersecurity, robotic surgery, image processing, virtual assistants, and clinical decision assistance [6, 7]. A computer that has been trained to think like a person is said to have AI. AI will advance clinical judgment and enhance patient diagnosis, prevention, and treatment.

In practice, no Artificial Intelligence tool will ever be able to take the place of a trained medical professional, but it will help them do their jobs better in the medical field. These evolving AI tools in medicine rely heavily on the availability of healthcare data. AI is a collection of technologies, not just one. Some of these technologies, such as machine learning, are widely used in healthcare. ML is a technique in which models are trained with historical data in order to correctly identify test inputs when presented with new data. ML is a kind of AI that is widely used [8]. Fig. (**1**) shows the various applications of AI in healthcare and pharmacy, which are diagnosis and treatment design; discovery and interactions with drugs; electronic health records; radiology; dermatology; psychological conditions primary care [9].

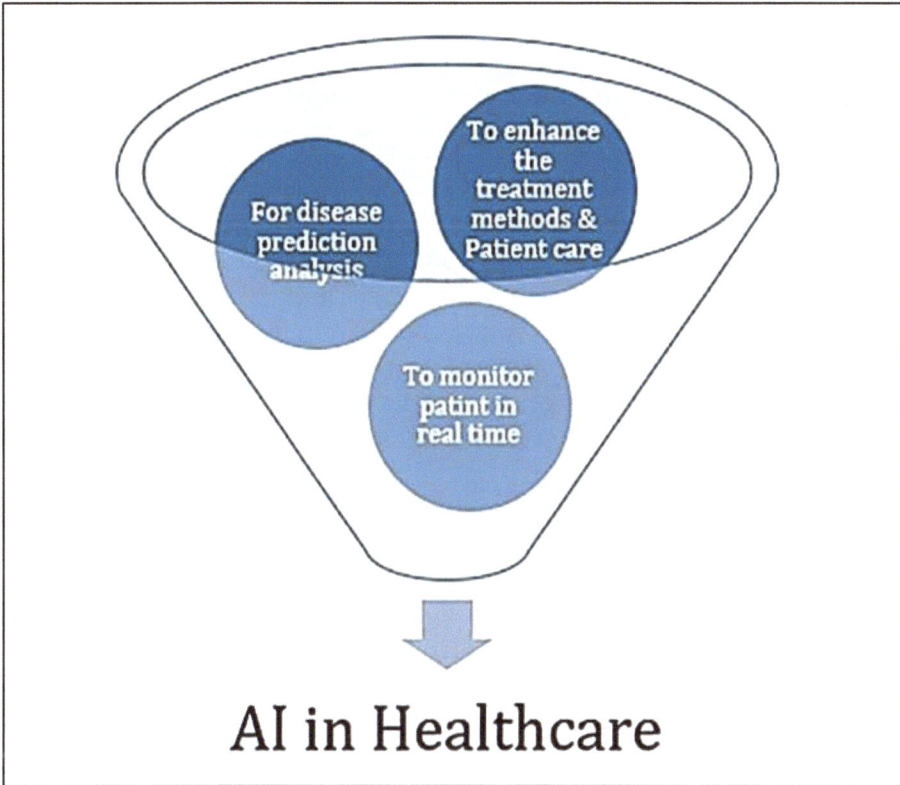

Fig. (1). AI applications in healthcare.

The most typical use of ML in healthcare is precision medicine.

AI in Precision Medicine

Precision medicine is a therapeutic area that has been established using information about a person's genetic makeup, way of life, gene expression, and environment. It's one of the most interesting and promising breakthroughs in contemporary medicine. It changes healthcare from being a universally applicable medical practice to the one that is data-driven and personalized, enabling more cost-effective spending and improved patient outcomes. It has aided in the treatment of numerous inflammatory diseases, including HIV, cardiovascular disease, and cancer.

Based on a patient's disease profile, reaction to a diagnosis or treatment, or prognostic data, precision medicine enables medical interventions for treating specific patients or groups of patients. Personalized treatment options consider variations in the genome, in addition to factors that contribute to medical treatment such as age, sex, geographic location, racial group, family medical history, immunological profile, microbiome, metabolic profile, and environmental vulnerability.

Precision medicine is more widely used for individual biology than for population biology throughout each and every stage of a patient's individual medical history. This entails gathering information from individuals, such as physiological tracking data, genetic information or EMR data, and using their processing for advanced models [10]. With the advent of precision medicine, it may be possible to substantially increase the traditional symptom-driven medicine with the combination of various multi-omics profiles with epidemiological, clinical, demographic and imaging data to create a diverse initial intake before a diagnosis is being developed and a more effective, efficient, and cost-effective personalized treatment is developed. The present focus of precision medicine is on gathering ever-increasing sample numbers, electronic health records, DNA sequence data including mobile health records, and other data. Fig. (**2**) depicts some of the algorithms which are commonly used in precision medicine: K-nearest neighbor (KNN), Support Vector Machine (SVM), random forest (RF), Discriminant Analysis, Naïve Bayes (NB), deep-learning model (DLM), logistic regression, genetic algorithm (GA), linear regression, decision tree (DT), and hidden Markov [11].

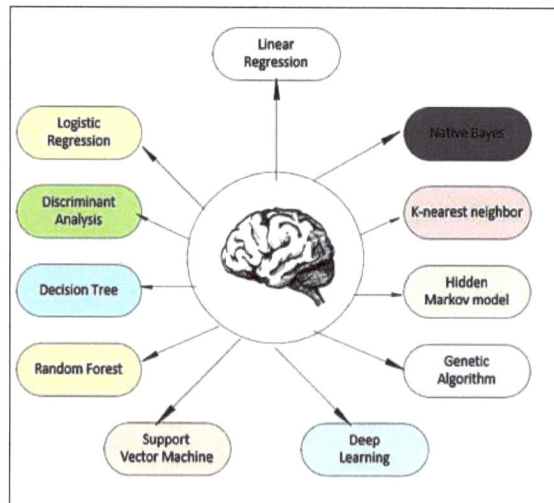

Fig. (2). leading machine learning algorithm in precision medicine [11].

Supervised, reinforcement learning and unsupervised learning are different types of ML. The main application of machine learning in the area of precision medicine is supervised learning, in which the result of the patient is known. Unsupervised learning makes use of collections of various uncategorized data to identify clusters or hidden patterns in the given data. The use of reinforcement learning in medicine is uncommon because it uses simulated data.

Big data is one of the main forces behind precision medicine. In the year 2011, US National Research Council created a comprehensive plan for advancing precision medicine; it is known as the creation of a "New Taxonomy" for various kinds of human diseases that would incorporate molecular, phenotypic and environmental data. Without analysis techniques to turn the data into knowledge that can be applied in therapeutic settings, big data alone have limited utility. Many academics are turning to AI to create reliable data analysis methods as a result of the advancements in parallel computing [12 - 14].

Cancer precision medicine has benefited from the use of functional genomic data. Functional genomics has recently made advancements in precision medicine for common non-cancer disorders including kidney disease. Below are some instances of how ML can be used in conjunction with other statistical methods to evaluate functional genomics data derived from disease-relevant tissues to help create precision medicine for prevalent non-cancer disorders [15, 16]. Using a machine learning algorithm, genomic data can be correlated with other data like metabolic profile, gut microbiota, and protein expression to determine a patient's personalized treatment also [17].

Application of AI in Critical Diseases

Data generated from various "omics," including proteomics, metabolomics, genomics and transcriptomics, can be combined using artificial intelligence. It has made it possible to describe almost any type of biomolecule, from DNA to metabolites, facilitating the study of complex biological systems. Searching for disease biomarkers using omics data facilitates the classification of patient subgroups and provides earlier diagnostic information about patients to improve patient management and can be very helpful in preventing any adverse outcomes. For the application of digital scans of lung tissue samples from the Cancer Genome Atlas, Coudray *et al*. have used the CNN in a very accurate and thorough way to detect the various types of lung cancer, including adenocarcinoma (LUAD), squamous cell carcinoma (LUSC) as well as normal lung tissue [18]. Table **1** depicts some of the algorithms which are commonly used in oncology.

Table 1. Algorithms of AI used in Cancer Diagnosis [19].

AI Algorithm Used	Oncology Studies
RF Regressor, Generic model, Wilcoxon Signed Ranked Test, Trans-issue Model, and Gene-Specific Model.	Breast and ovarian cancer [20, 21].
RF and CNN.	Kidney cancer [21]
SVM, genetic algorithm ROC, ELM, Two-stage Fuzzy Neural Network and Neural Networks	Prostate cancer [21 - 23]
Ensemble model LDA, SVM, Decision Tree, KNN, ROC and ANN, SVM, RPART, Semi-Supervised Learning, Neural Networks, ANN, Calibration Slope, Logistic Regression, RF and Generalized Boosted Model	Breast cancer [24 - 27]
C4.5, SVM, RF, Naive Bayes, PLS-DA, KNN, LASSO	Colonic cancer [28]
R89-restricted Boltzmann Machine, Pathway Based Deep Clustering Model and Deep Belief Network	GBM and ovarian cancer [29]
SVM, Cox Hazard Regression Model, Log-Rank Test and GA	Ovarian cancer [30]
KNN, SVM, Bagged SVM, RF, Adaboost and GBT	Bladder cancer [31]
RUSBoost, DT, Matthews Correlation Coefficient and Adaboost, algorithm	Gliomas [32]
Linear Regression, SVM, Gradient Boosting Machines and Decision Tree	Lung cancer [33]

Patients' cardiovascular disorders can be diagnosed using Artificial Intelligence. Chest radiographs can be used to identify congestive heart failure using a neural network classifier. In ML, electro cardiology is most likely the most advanced field. Computers can now extract additional data from ECG records that are inaccessible to human interpretation thanks to data mining. Convolutional neural networks had a great predictive performance (AUC of 90% after internal validation) when it came to detecting atrial fibrillation in patients using ECGs taken in sinus rhythm [34].

AI in Drug Design

Drug design and development is a time-consuming and expensive process. Deep learning (DL) is now the focus of contemporary AI applications in drug design. Over the past ten years, deep generative models (DGM) and graph neural networks (GNN) have been developed, in which DGM was first created for other zones including natural language processing (NLP) and computer vision, and has dominated the field of molecular design and synthesis of drugs [35]. Stokes *et al.* (2020) used a GNN to identify a compound library that allowed them to find the antibiotic halicin, which is useful for treating infections in animal models [36]. In behavior experiments using GNN, Sakai *et al.* (2021) discovered antidepressant properties of a particularly potent serotonin transporter inhibitor [37].

Recent AI models including generative adversarial networks (GAN), GNN, recurrent neural networks (RNN), variation autoencoders (VAE), flow, and reinforcement learning (RL) have seen growth in the last five years in drug development applications [38].

ARTIFICIAL INTELLIGENCE AND MEDICAL VISUALIZATION

As the complexity of medical knowledge grows, electronic clinical decision support in healthcare becomes increasingly important and the use of artificial intelligence becomes more promising [39]. AI algorithms require enormous volumes of data to work properly, and as Picture Archiving and Communication Systems (PACS) are being widely used, imaging data is growing quickly. Through the use of machine learning technology and for clinical applications, these data have been successfully combined, however, the reach of these algorithms to various healthcare visualization systems is still limited [40].

It can be difficult to interpret data which comes in the form of an image source or in the form of a video clip. To develop the ability to recognize medical events, individuals in the field must go for training for many years to become field experts. In addition, there is a need to constantly update their knowledge as new findings and data become accessible. However, there is a severe lack of subject matter specialists. As a result, a novel strategy is required, and AI holds promise as the tool to close this demand gap.

Machine Vision in the Field of Diagnosis and Surgery

Since statistical signal processing has been the basic foundation for computer vision, artificial neural networks are currently being used more frequently as the preferred learning method. Here, computer vision algorithms (CVA) are used for categorizing the pictures of lesions in the skin and other tissues developed by using DLM. Video data could offer a greater data value based on the resolution of video clips over time because it is predicted to include 25 times as much data as higher-resolution diagnostic images like CT. Although it is still early, video analysis offers a lot of potential for clinical decision assistance. As an illustration, real-time video analysis of a laparoscopic surgery showed 92.8 percent accuracy in step recognition as well as the startling discovery of missing stages [41].

One significant use of computer vision and AI in surgical technology is to improve certain procedures and abilities like suturing and knot-tying. In various surgical operations, such as animal bowel anastomosis, the smart tissue autonomous robot (STAR) at Johns Hopkins University has demonstrated that it is on par with human surgeons in terms of effectiveness. Although a fully automatic robotic surgeon is still a dream for many years to come, academics are interested in applying AI to improve several elements of surgery. For example,- Endoscopic surgery videos are used for training medical professionals, analyzing medical research, and documenting routine clinical practice [42]. Surgeons need to actively participate in the development of such tools that are clinically relevant and of good quality so that the translation of obtained tools from the lab to the clinical sector can be facilitated.

Medical Image Analysis using DL

Deep learning-based medical image analysis is a rapidly growing field of study. Since its introduction, DLA has been an integral part of medical imaging for diagnosing disease [43]. The word "deep" may be used to refer to the multi-layered nature of ML, and of all deep learning technologies, CNN has been shown to be the most valuable in the area of image recognition. CNNs are widely influenced by human visual cortex for example radiologists must have knowledge and ability to learn continuously by correlating and associating the obtained interpretation of radiographic images from x-ray machines with ground truth during medical education. Many features of an image may be identified by Image recognition performed in the human visual cortex. In addition, in the form of medical images, CNN requires a large amount of training data along with labels about what the images should be. At each hidden training level, the CNN can accumulate the weights and filters applied (features of the area in the image) for the improvement of the performance given in training data [44].

Virtual Reality Augmented Reality (AR) in the Healthcare Space

The integration or addition of digital data into the user's physical environment is known as augmented reality. On top of the Real Environment, computer visuals are superimposed. With this technology, users can interact with digital items that have been placed in the real world [45]. In order to provide users of virtual reality (VR) a sense of the physical world, some aspects and materials from the real world are added to the virtual environment. Nearly every sector is getting influenced nowadays by how AR and VR are altering our perception and interaction with the environment. With the help of these technologies, 3D immersive displays and anatomical comprehension are now achievable. Medical applications are extensive and have an impact on all aspects of medical care, including surgical technique and learning gross anatomy, patient-specific pre- and post-operative planning, intraoperative guidance, and diagnostic and treatment modalities in psychology, rehabilitation, and pain management [46].

AR and VR may be integrated into all phases of the healthcare system. The healthcare systems may be implemented for medical students in the early stages of education, professionally trained students, and experienced surgeons. The current education system is limited and this can be an obstacle in areas such as medicine. Medicine can be kept or shown in the form of art, and future clinicians are artists. These individuals need to have specific skills to meet the needs of an ever-evolving profession. At the beginning of medical school, students are taught various concepts without experiencing them in actual life. So, technologies like VR and AR may improve the learning experience in the field of future medical disciplines and in the field of health-related disciplines by making the tasks and procedures more game-like. Novel and complex surgical procedures can be offered and taught to medical students, or anatomy can be learned through AR without the need to engage with various actual patients in the preliminary stage or perform autopsies on real cadavers. Eventually, these students will work with actual patients, but right now, the goal is to get them trained early on in order to save future training expenses [44].

A recent study found that two groups of surgical trainees performed a distinct procedure of Mastoidectomy – the first group underwent standard training, while the other group trained on a virtual reality simulator. After completing training using virtual reality, surgical dissection improved significantly among those participants. In contrast, real-life performance may be better achieved with augmented reality technologies [47]. One of the interesting ways that may help to focus on the task without ever being distracted will be by moving the visual field away from the region of interest by wearing lightweight headsets (like Google Glass or Microsoft HoloLens) that the videos or images related to the area of interest.

Patient Experience

People use audiovisual cues to interact with the surrounding environment and use their limbs to move around in this world. A recent study found that immersive VR is helping stroke patients recover more quickly [48]. These patients may find this immersive experience helpful in their rehabilitation, as it engages multiple times a day, their upper limb movement. This may help to create neuroplasticity with the subtle return of normal motor function in these regions. These technologies may aid people suffering from cancer or mental health conditions to deal with the pain and illness that these conditions can cause. A study has found that terminal cancer patients may benefit using this technology while suffering minimal physical discomfort and, therefore, improved states of relaxation, entertainment, and well-being [49].

Intelligent Personalized Health Records

Personalized health records (PHR) are one of the main factors that transform health care all over the world. PHR can be either physical or electronic as they become more common in the future. This includes all self-reports, including health issues and treatments, vital signs and activity, and nutritional data such as nutritional content and caloric intake records recorded using personal devices such as smartphones and smartwatches. PHR would have traditionally been inclined toward the needs of physicians and typically not included features that are specific to the needs of patients. Individual health records should be implemented to promote self-management and improve patient outcomes. It aims to give patients full control over their health while freeing up time for doctors to focus on more pressing issues [44].

Wearables and Health Monitoring

People have been using doctors as their primary source of information regarding their health for quite some time, and this trend is never going to end . The constantly expanding field of wearable technology, however, is altering this norm. The key to the early adoption and success of WHDs, a new technology that enables continuous measurement and diagnosis of certain vital signs under different conditions, is the application flexibility that enables users to now track their activity while running, walking, or even swimming in the water. The aim is to empower every individual by allowing them to understand their own health data and make decisions about their care. WHDs empower individuals by providing them with the opportunity to take control of their lives (Fig. **3**) [44].

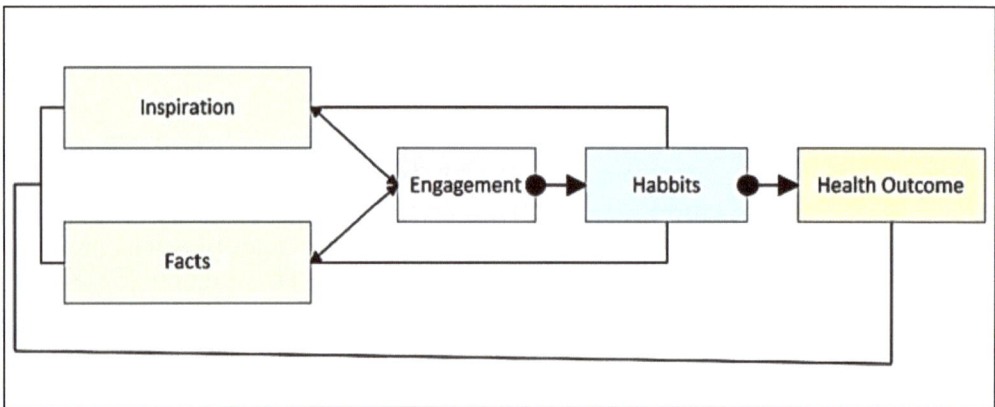

Fig. (3). Simple, interrelated criteria determine patient health outcome [44].

At first glance, wearable devices seem like regular bracelets and watches. However, these tools have the potential to narrow the chasm that exists across many scientific fields like electronic engineering, biomedical engineering, computer programming, materials and data science. Remote monitoring and detection of early signs of illness can be of great benefit to people with chronic patients and the elderly. By wearing a smart device or manually entering data for a long time, you can communicate with healthcare professionals without disturbing your daily life [50].

ARTIFICIAL INTELLIGENCE-POWERED DEVICES AND ROBOTICS

Robots are being used in many different aspects of healthcare, from performing surgical procedures to assisting people with disabilities, to helping doctors with their work. Most current robot development efforts include some form of artificial intelligence for improved performance in classifications, language recognition, image processing, and more. From diagnosing COVID-19 to stopping its spread, and from reducing the workloads of health workers to speed up communication operations, AI and robotics technologies have effectively matched the objectives of public health. The use of robotics and AI technologies development has helped substantially in the invention of the COVID vaccine in the vaccination process as well. To process the stream of information on the most recent discoveries and inquiry findings, a number of AI-enabled big data analysis techniques are employed. The administration and distribution of vaccines has made excellent use of UAV and RPA platforms. For any tragedy, the solutions provided by technological fields to the recurring problems of this pandemic can serve as an excellent model [51].

Robots Enabled Healthcare

As the demand for doctors continues to grow, it has become necessary for elderly patients to provide personalized healthcare to those with incurable diseases, as well as to take appropriate measures during emergency circumstances for older patients. The healthcare industry is experimenting with various changes in the use of robots to deliver personalized healthcare to patients according to their needs [52].

One of the common features of healthcare robots is their ability to monitor the health of their users. Robots and artificial intelligence can use tomeasure the vital signs of patients using different types of sensors and the resulting data can be shared with the doctors for consultation and proper treatment. Fig. (**4**) shows various types of sensors used for collecting patients' health data. Robots and artificial intelligence can also be integrated with various communication capabilities allowing patients to consult with the doctor directly through video calls. Robots must have the ability and adaptability to merge all of the existing technologies, making the interface simple and easy to use. There are a variety of existing robots used for health monitoring, such as Pearl, Hector, and Cafero. There are also robots developed to monitor vital signs only, such as Hopis [53 - 56].

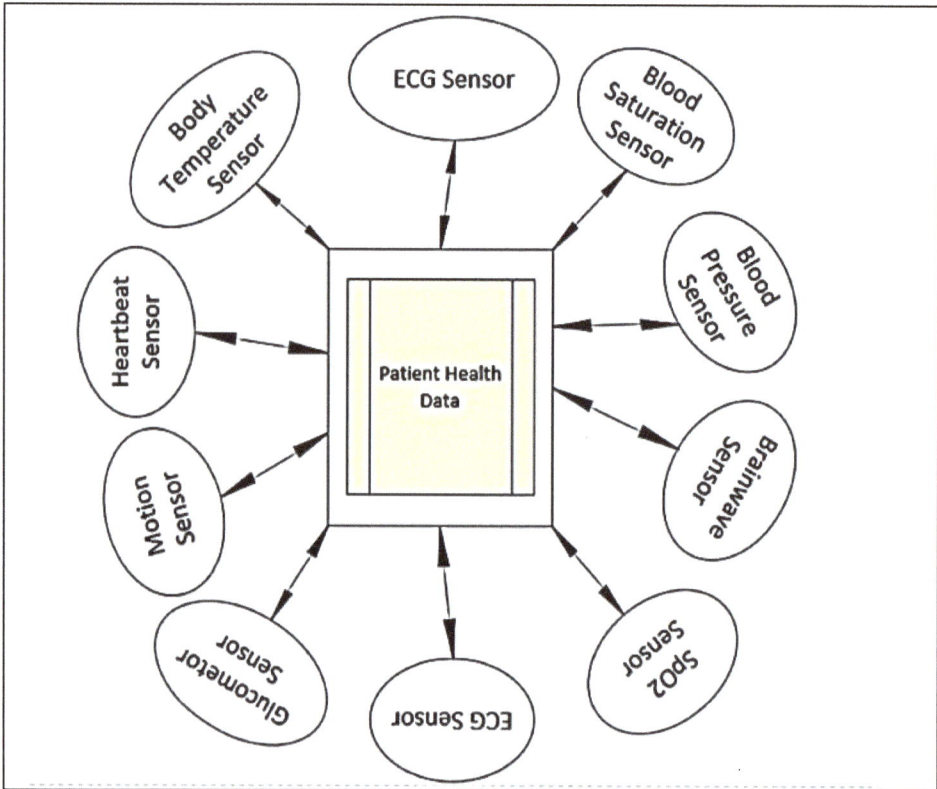

Fig. (4). Various sensors used for the collection of patient health data [52].

It is important that healthcare robots under development consider the needs of the elderly, as well as the needs of the whole community. In the near future, it is possible that robotics may become prominent as one of the most crucial advantageous factors in creating and developing intelligent digital worlds. Telecommunication robots can contact healthcare providers, doctors as well as family members and friends of respective patients. With the advancement of technology, robots are being developed as social companions to help patients or users feel less lonely [52]. Robots like Paro, AIBO, and others are typically developed to provide fellowship for old age people and serve as companions that can interact with patients and help them to take care of themselves [57, 58].

The elderly often experiences a wide range of problems with their health and cognitive issues, which makes it important to monitor their physical as well as mental health and to manage their medical issues appropriately in order to assist them to live independently. Even if older people are able to cope with the challenges of ageing, they may end up feeling overwhelmed and alone. The multi-functional robots are being synthesized and developed to prevent interventions and store a large amount of data related to medical, academic or other related sectors. If there are any emergencies, they will also alert an ambulance [52].

Ambient Assisted Living

As the aged population are living with chronic disorders throughout their lives. However, most people want to live independently into their old age. Data show that half of the people of age 65 or more have some kind of disorder, which is more than 350 lakhs in the United States alone itself. Most of the people want to remain autonomous and maintain their control even in old age [59]. Industries, governments and various companies are advertising the concept of "independent living" – a way of living in which people can live independently in their own homes. AAL includes promoting healthy lifestyles for vulnerable groups, increasing autonomy and mobility in older adults, increasing safety, support, and productivity, and ultimately living in the environment of their choice. There are multiple goals, such as helping to improve the life quality of human beings. Ambient-assisted living applications are typically used to collect data through cameras, and sensors and use various artificial intelligence tools to develop intelligent systems [60]. One way to implement AAL is to use a smart home or assistive robot.

IoT Integrated Smart Home

A smart home is just like a conventional one, with the addition of various sensors, actuators and tracking tools that can make it "smart" and provide easier lives to the residents of such homes. Remote tracking, reminders, alert creation, behavioural analytics, and robot assistance are some of the other common ambient-assisted living apps that can be integrated into a smart home or utilised independently [52]. There are several studies investigating the usefulness of smart homes for people with dementia. These homes can make lives more comfortable for those with a disease condition, by facilitating in communication and safety. In the market, various expensive sensors used in an Internet of Things (IoT) architecture may be helpful to detect unusual behavior in any home. For example, sensors can be set in different parts of the home like the kitchen, bedroom and bathroom to ensure the security of a particular home. Sensors can also be placed on the stove for the detection of the use of the gas stove and notify the patient if the gas stove is not off after use. In bathroom applications, a lamp sensor and a bathtub sensor can be attached to provide a signal if it remains left on, to the elderly person living in a smart home [61].

The sensors send data or signals to a nearby communication or computing system that can be processed using machine learning algorithms, or uploaded to the cloud for further processing. If necessary, the sensors can also send alert signals to relatives or healthcare professionals (Fig. **5**).

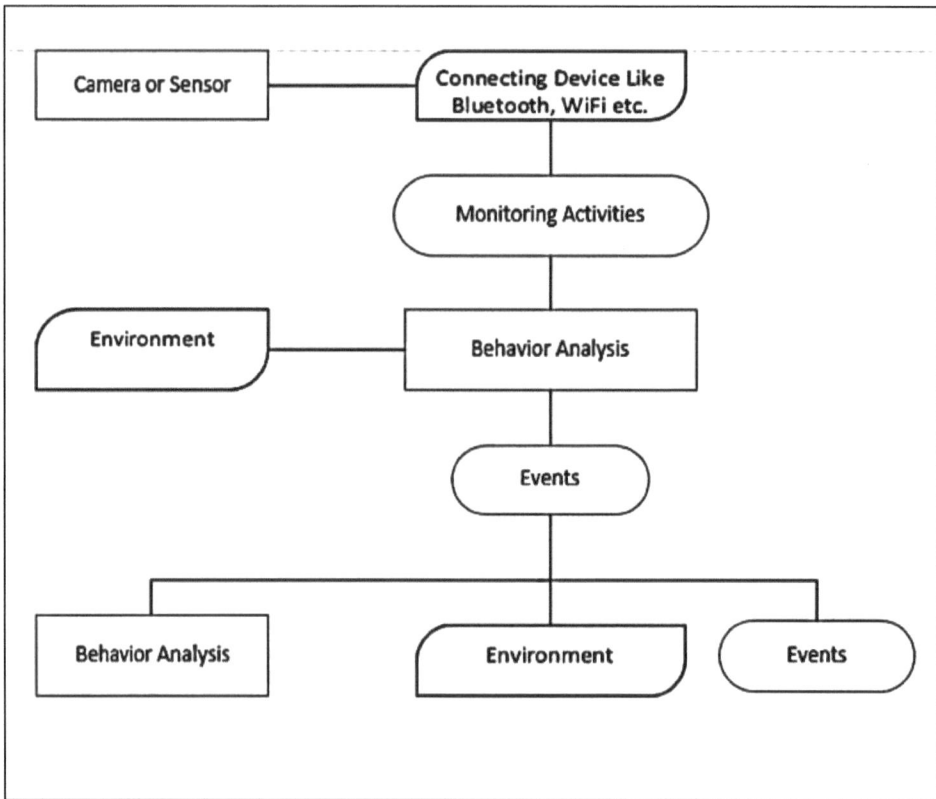

Fig. (5). A typical smart home process diagram [52].

Cognitive Assistants

Many older people in this world experience and talk about cognitive decline, staying alert, accessing memory, and struggling with problem-solving tasks. Cognitive stimulation is a common type of approach to rehabilitation followed after multiple sclerosis, brain injuries, or trauma. Virtrael is a well-known cognitive stimulation platform that helps assess, train, and stimulate different kinds of cognitive skills that can be declining in patients. Virtrael's three primary important functions—configuration, games, and communication—carry out the ongoing effort, which is based on memory training to recall information. An administrator can assign a therapist to a patient in the configuration mode so that the latter can be briefed about the treatment plan. The patient can talk to their therapist and other patients *via* the communication tool.

Memory, focus, and planning are just some of the cognitive abilities that the games will actively aim to cultivate [62].

BENEFITS OF AI IN HEALTHCARE

The use of AI in healthcare has the potential for both beneficial and detrimental outcomes. The extensive use of modern AI technology in the healthcare system has many benefits, like helping patients to make better decisions, and use better technology to get better results, as well as other benefits such as reducing referrals, reducing checkup and treatment time, saving resources, and saving money. It can improve the medical facilities in remote and rural areas and it can also help in improving the recruitment and retention in such areas. Finally, up to some extent, this may help make the world's healthcare system more efficient and the challenges facilitate sustainable implementation and early adoption in the health system. AI can integrate patient records and summarize physicians' health concerns. Instead of manually searching patient data, AI can search for it at a much faster rate through available data and also highlights the key details in the data [63]. AI can also reduce the workload in the areas dealing with diagnostic images and their interpretation. In consideration of the enormous workload and stress of healthcare workers, it seems natural to commit mistakes. but, in healthcare activities, mistakes can lead to devastating consequences. Therefore, it is optimal if AI acts as a second eye to reduce mistakes by using its precise algorithm. AI can also be used to decrease hospitalization cases by avoiding any nonessential cases to occur [64].

Some tasks that would typically be carried out by medical experts can be replaced by AI. Many administrative tasks performed by doctors or nurses are repetitive in nature and required the little awareness, which can be easily replaced by AI applications [65, 66]. AI holds considerable promise for enhancing healthcare procedures and patient care. Here AI can not only complement the work of medical professionals but also expand their sphere of action [67]. AI can also serve as a decision-making tool. From a user perspective, the technology cannot be used optimally but there is a need to adopt artificial intelligence in the public or government health sector. Adopting AI in the public health sector provides a blueprint for where and how this type of research should be focused in other sectors of the public sector [68].

LIMITATION & FUTURE SCOPE OF AI IN HEALTH SERVICE

While the multidisciplinary nature of AI demonstrates to be a revolution in technology and an accelerator for complex jobs, it also comes with a lot of highly contested challenges in terms of the ethics and the law. Although AI applications have the tendency to eliminate human bias and mistakes, biases in the training data can potentially reflect and perpetuate prejudices in the applications [63]. Training the AI model will require a large amount of input about your health data or other relevant information. This bias can arise when the data and information used for instructing does not represent the target population, and when the data and information used to train the AI model are insufficient or incomplete due to social discrimination, *etc* [69]. The data relating to a patient's health services are the most sensitive bits of information that an individual can retain about another person. Because privacy is connected to patient autonomy or self-governance, personal identity, and well-being, the ethical principle of respecting an individual's right to privacy is of the utmost importance in the field of healthcare [70]. Therefore, it is morally required to protect patient privacy and implement rigorous procedures for obtaining informed permission. Given that science is a retreat, some discoveries can eventually lead to damage. In light of this, it is essential to give due consideration to dual ethical standards in applying AI, such as in stem cell research and gene editing. Using AI in healthcare applications requires adherence to biomedical ethical guidelines, as is the case with any new scientific approach. The four values are self-control, self-interest, non-crime, and justice. The principles of consent, privacy, and safety should be considered and practiced when implementing a system [71, 72].

Despite significant progress in the past several years, there are still many hurdles in the way of reliable clinical diagnostics that need to be overcome in order to properly treat new ailments and diseases. Even medical practitioners are aware of the difficulties that need to be overcome before illness may be recognized using AI.

REFERENCES

[1]   D.D. Miller, and E.W. Brown, "Artificial Intelligence in Medical Practice: The Question to the Answer?", *Am. J. Med.,* vol. 131, no. 2, pp. 129-133, 2018.
[http://dx.doi.org/10.1016/j.amjmed.2017.10.035] [PMID: 29126825]

[2]   L.G. Pee, S.L. Pan, and L. Cui, "Artificial intelligence in healthcare robots: A social informatics study of knowledge embodiment", *J. Assoc. Inf. Sci. Technol.,* vol. 70, no. 4, pp. 351-369, 2019.
[http://dx.doi.org/10.1002/asi.24145]

[3]   J. Bresnick, "Artificial Intelligence in Healthcare Market to See 40% CAGR Surge", Available at: https://healthitanalytics.com/news/artificial-intelligence-in-healthcare-market-to-see-40-cagr-surge (2017).

[4]   S.P. Yadav, and S. Yadav, "Fusion of Medical Images using a Wavelet Methodology: A Survey", In: *In: IEIE Transactions on Smart Processing & Computing* vol. 8. The Institute of Electronics Engineers: Korea, 2019, no. 4, pp. 265-271.
[http://dx.doi.org/10.5573/IEIESPC.2019.8.4.265]

[5]   D. Lee, and S.N. Yoon, "Application of Artificial Intelligence-Based Technologies in the Healthcare Industry: Opportunities and Challenges", *Int. J. Environ. Res. Public Health,* vol. 18, no. 1, p. 271, 2021.
[http://dx.doi.org/10.3390/ijerph18010271] [PMID: 33401373]

[6]   B. Marr, "How Is AI Used In Healthcare - 5 Powerful Real-World Examples That Show The Latest Advances", Available at: https://www.forbes.com/sites/bernardmarr/2018/07/27/how-is-ai-use--in-healthcare-5-powerful-real-world-examples-that-show-the-latest-advances/?sh=c66d26c5dfbe (2018).

[7]   "10 Promising AI Applications in Health Care", Available at: https://www.investkl.gov.my/clients/ asset_28B5D799-69B3-4BCB-B61B-D284619547A3/uploads/10-Promising-AI-Applications-in-HealthCare.PDF(2018).

[8]   S.P. Yadav, and S. Yadav, "Fusion of Medical Images in Wavelet Domain: A Discrete Mathematical Model", In: *In Ingeniería Solidaria.* vol. 14. Universidad Cooperativa de Colombia: UCC, 2018, no. 25, pp. 1-11.
[http://dx.doi.org/10.16925/.v14i0.2236]

[9]   A. Eren, A. Subasi, and O. Coskun, "A decision support system for telemedicine through the mobile telecommunications platform", *J. Med. Syst.,* vol. 32, no. 1, pp. 31-35, 2008.
[http://dx.doi.org/10.1007/s10916-007-9104-x] [PMID: 18333403]

[10]   L. Konieczny, and I. Roterman, "Personalized precision medicine", *Bio-Algorithms and Med-Systems,* vol. 15, no. 4, p. 47, 2019.
[http://dx.doi.org/10.1515/bams-2019-0047]

[11]   S. Quazi, "Role of Artificial Intelligence and Machine Learning in Bioinformatics: Drug Discovery and Drug Repurposing", *Preprints (Basel),* .
[http://dx.doi.org/10.20944/preprints202105.0346.v1]

[12]   *Toward Precision Medicine: Building a Knowledge Network for Biomedical Research and a New Taxonomy of Disease.* National Academies Press: Washington, D.C., 2011.
[http://dx.doi.org/10.17226/13284] [PMID: 22536618]

[13]   J.S. Beckmann, and D. Lew, "Reconciling evidence-based medicine and precision medicine in the era of big data: challenges and opportunities", *Genome Med.,* vol. 8, no. 1, p. 134, 2016.
[http://dx.doi.org/10.1186/s13073-016-0388-7] [PMID: 27993174]

[14]   S. P. Yadav, and S. Yadav, "Mathematical implementation of fusion of medical images in continuous wavelet domain", *Journal of Advanced Research in dynamical and control system,* vol. 10, no. 10, pp. 45-54, 2019.

[15]   C. Huang, R. Mezencev, J.F. McDonald, and F. Vannberg, "Open source machine-learning algorithms for the prediction of optimal cancer drug therapies", *PLoS One,* vol. 12, no. 10, p. e0186906, 2017.
[http://dx.doi.org/10.1371/journal.pone.0186906] [PMID: 29073279]

[16]   S. Huang, N. Cai, P.P. Pacheco, S. Narrandes, Y. Wang, and W. Xu, "Applications of Support Vector Machine (SVM) Learning in Cancer Genomics", *Cancer Genomics Proteomics,* vol. 15, no. 1, pp. 41-51, 2018.
[http://dx.doi.org/10.21873/cgp.20063] [PMID: 29275361]

[17]   J. Love-Koh, A. Peel, J.C. Rejon-Parrilla, K. Ennis, R. Lovett, A. Manca, A. Chalkidou, H. Wood, and M. Taylor, "The Future of Precision Medicine: Potential Impacts for Health Technology Assessment", *PharmacoEconomics,* vol. 36, no. 12, pp. 1439-1451, 2018.
[http://dx.doi.org/10.1007/s40273-018-0686-6] [PMID: 30003435]

[18]   N. Coudray, P.S. Ocampo, T. Sakellaropoulos, N. Narula, M. Snuderl, D. Fenyö, A.L. Moreira, N. Razavian, and A. Tsirigos, "Classification and mutation prediction from non–small cell lung cancer histopathology images using deep learning", *Nat. Med.,* vol. 24, no. 10, pp. 1559-1567, 2018.
[http://dx.doi.org/10.1038/s41591-018-0177-5] [PMID: 30224757]

[19]   S. Quazi, "Artificial intelligence and machine learning in precision and genomic medicine", *Med. Oncol.,* vol. 39, no. 8, p. 120, 2022.
[http://dx.doi.org/10.1007/s12032-022-01711-1] [PMID: 35704152]

[20]   F. Azuaje, S.Y. Kim, D. Perez Hernandez, and G. Dittmar, "Connecting Histopathology Imaging and Proteomics in Kidney Cancer through Machine Learning", *J. Clin. Med.,* vol. 8, no. 10, p. 1535, 2019.
[http://dx.doi.org/10.3390/jcm8101535] [PMID: 31557788]

[21]   S. Zhang, Y. Xu, X. Hui, F. Yang, Y. Hu, J. Shao, H. Liang, and Y. Wang, "Improvement in prediction of prostate cancer prognosis with somatic mutational signatures", *J. Cancer,* vol. 8, no. 16, pp. 3261-3267, 2017.
[http://dx.doi.org/10.7150/jca.21261] [PMID: 29158798]

[22]   P. S, D. P. Mahato, and N. T. D. Linh, *Distributed Artificial Intelligence.,* S.P. Yadav, D.P. Mahato, N. T. D. Linh,, Eds., 1st Edition. CRC Press.: Boca Raton, 2020, p. 336.
[http://dx.doi.org/10.1201/9781003038467]

[23]   S. Jović, M. Miljković, M. Ivanović, M. Šaranović, and M. Arsić, "Prostate Cancer Probability Prediction By Machine Learning Technique", *Cancer Invest.,* vol. 35, no. 10, pp. 647-651, 2017.
[http://dx.doi.org/10.1080/07357907.2017.1406496] [PMID: 29243988]

[24]   M. Zhao, Y. Tang, H. Kim, and K. Hasegawa, "Machine Learning With K-Means Dimensional Reduction for Predicting Survival Outcomes in Patients With Breast Cancer", *Cancer Inform.,* 2018.
[http://dx.doi.org/10.1177/1176935118810215] [PMID: 30455569]

[25]   F.M. Alakwaa, K. Chaudhary, and L.X. Garmire, "Deep Learning Accurately Predicts Estrogen Receptor Status in Breast Cancer Metabolomics Data", *J. Proteome Res.,* vol. 17, no. 1, pp. 337-347, 2018.
[http://dx.doi.org/10.1021/acs.jproteome.7b00595] [PMID: 29110491]

[26]   K. Park, A. Ali, D. Kim, Y. An, M. Kim, and H. Shin, "Robust predictive model for evaluating breast cancer survivability", *Eng. Appl. Artif. Intell.,* vol. 26, no. 9, pp. 2194-2205, 2013.
[http://dx.doi.org/10.1016/j.engappai.2013.06.013]

[27]   D. Delen, G. Walker, and A. Kadam, "Predicting breast cancer survivability: a comparison of three data mining methods", *Artif. Intell. Med.,* vol. 34, no. 2, pp. 113-127, 2005.
[http://dx.doi.org/10.1016/j.artmed.2004.07.002] [PMID: 15894176]

[28]   R. Eisner, R. Greiner, V. Tso, H. Wang, and R.N. Fedorak, "A machine-learned predictor of colonic polyps based on urinary metabolomics", *BioMed Res. Int.,* vol. 2013, pp. 1-11, 2013.
[http://dx.doi.org/10.1155/2013/303982] [PMID: 24307992]

[29]   T. Mallavarapu, J. Hao, Y. Kim, J.H. Oh, and M. Kang, "Pathway-based deep clustering for molecular subtyping of cancer", *Methods,* vol. 173, pp. 24-31, 2020.
[http://dx.doi.org/10.1016/j.ymeth.2019.06.017] [PMID: 31247294]

[30]   T.P. Lu, K.T. Kuo, C.H. Chen, M.C. Chang, H.P. Lin, Y.H. Hu, Y.C. Chiang, W.F. Cheng, and C.A. Chen, "Developing a Prognostic Gene Panel of Epithelial Ovarian Cancer Patients by a Machine Learning Model", *Cancers (Basel),* vol. 11, no. 2, p. 270, 2019.
[http://dx.doi.org/10.3390/cancers11020270] [PMID: 30823599]

[31]   Z. Hasnain, J. Mason, K. Gill, G. Miranda, I.S. Gill, P. Kuhn, and P.K. Newton, "Machine learning models for predicting post-cystectomy recurrence and survival in bladder cancer patients", *PLoS One,* vol. 14, no. 2, p. e0210976, 2019.
[http://dx.doi.org/10.1371/journal.pone.0210976] [PMID: 30785915]

[32]   C.F. Lu, F.T. Hsu, K.L.C. Hsieh, Y.C.J. Kao, S.J. Cheng, J.B.K. Hsu, P.H. Tsai, R.J. Chen, C.C. Huang, Y. Yen, and C.Y. Chen, "Machine Learning–Based Radiomics for Molecular Subtyping of Gliomas", *Clin. Cancer Res.,* vol. 24, no. 18, pp. 4429-4436, 2018.
[http://dx.doi.org/10.1158/1078-0432.CCR-17-3445] [PMID: 29789422]

[33]   C.M. Lynch, B. Abdollahi, J.D. Fuqua, A.R. de Carlo, J.A. Bartholomai, R.N. Balgemann, V.H. van Berkel, and H.B. Frieboes, "Prediction of lung cancer patient survival via supervised machine learning classification techniques", *Int. J. Med. Inform.,* vol. 108, pp. 1-8, 2017.
[http://dx.doi.org/10.1016/j.ijmedinf.2017.09.013] [PMID: 29132615]

[34]   Z.I. Attia, P.A. Noseworthy, F. Lopez-Jimenez, S.J. Asirvatham, A.J. Deshmukh, B.J. Gersh, R.E. Carter, X. Yao, A.A. Rabinstein, B.J. Erickson, S. Kapa, and P.A. Friedman, "An artificial intelligence-enabled ECG algorithm for the identification of patients with atrial fibrillation during sinus rhythm: a retrospective analysis of outcome prediction", *Lancet,* vol. 394, no. 10201, pp. 861-867, 2019.
[http://dx.doi.org/10.1016/S0140-6736(19)31721-0] [PMID: 31378392]

[35]   A. Oussidi, and A. Elhassouny, "Deep generative models: Survey", *2018 International Conference on Intelligent Systems and Computer Vision (ISCV),* pp. 1-8, 2018.
[http://dx.doi.org/10.1109/ISACV.2018.8354080]

[36]   J.M. Stokes, K. Yang, K. Swanson, W. Jin, A. Cubillos-Ruiz, N.M. Donghia, C.R. MacNair, S. French, L.A. Carfrae, Z. Bloom-Ackermann, V.M. Tran, A. Chiappino-Pepe, A.H. Badran, I.W. Andrews, E.J. Chory, G.M. Church, E.D. Brown, T.S. Jaakkola, R. Barzilay, and J.J. Collins, "A Deep Learning Approach to Antibiotic Discovery", *Cell,* vol. 180, no. 4, pp. 688-702.e13, 2020.
[http://dx.doi.org/10.1016/j.cell.2020.01.021] [PMID: 32084340]

[37]   M. Sakai, K. Nagayasu, N. Shibui, C. Andoh, K. Takayama, H. Shirakawa, and S. Kaneko, "Prediction of pharmacological activities from chemical structures with graph convolutional neural networks", *Sci. Rep.,* vol. 11, no. 1, p. 525, 2021.
[http://dx.doi.org/10.1038/s41598-020-80113-7] [PMID: 33436854]

[38]   Y. Zhang, "An In-depth Summary of Recent Artificial Intelligence Applications Drug Design", 2021.
[http://dx.doi.org/10.48550/arXiv.2110.05478.]

[39]   P. Nadkarni, and S.A. Merchant, "Enhancing medical-imaging artificial intelligence through holistic use of time-tested key imaging and clinical parameters: Future insights", *Artificial Intelligence in Medical Imaging,* vol. 3, no. 3, pp. 55-69, 2022.
[http://dx.doi.org/10.35711/aimi.v3.i3.55]

[40]   M.R. Fromherz, and M.S. Makary, "Artificial intelligence: Advances and new frontiers in medical imaging", *Artificial Intelligence in Medical Imaging,* vol. 3, no. 2, pp. 33-41, 2022.
[http://dx.doi.org/10.35711/aimi.v3.i2.33]

[41]   D.A. Hashimoto, G. Rosman, D. Rus, and O.R. Meireles, "Artificial Intelligence in Surgery: Promises and Perils", *Ann. Surg.,* vol. 268, no. 1, pp. 70-76, 2018.
[http://dx.doi.org/10.1097/SLA.0000000000002693] [PMID: 29389679]

[42]   S. Petscharnig, and K. Schöffmann, "Learning laparoscopic video shot classification for gynecological surgery", *Multimedia Tools Appl.,* vol. 77, no. 7, pp. 8061-8079, 2018.
[http://dx.doi.org/10.1007/s11042-017-4699-5]

[43]   M. Puttagunta, and S. Ravi, "Medical image analysis based on deep learning approach", *Multimedia Tools Appl.,* vol. 80, no. 16, pp. 24365-24398, 2021.
[http://dx.doi.org/10.1007/s11042-021-10707-4] [PMID: 33841033]

[44]   A. Bohr, and K. Memarzadeh, "The rise of artificial intelligence in healthcare applications", In: *Artificial Intelligence in Healthcare.* Elsevier, 2020, pp. 25-60.
[http://dx.doi.org/10.1016/B978-0-12-818438-7.00002-2]

[45]   K. Sethiya, "Augmented Reality (AR) in Healthcare", *J. Interdiscipl. Cycle Res.,* vol. XII, no. XI, pp. 343-359, 2020.
[http://dx.doi.org/18.0002.JICR.2020.V12I11.008301.317122134]

[46]   S. S. S. Wayne, L Monsky, and James Ryan, "Virtual and Augmented Reality Applications in Medicine and Surgery-The Fantastic Voyage is here", *Natomy & Physiology: Current Research*, vol. 9, 2019, no. 1. https://www.longdom.org/open-access/virtual-and-augmented-reality-applications- in-med-icine-and-surgerythe-fantastic-voyage-is-here-25534.html

[47]   M. Frendø, L. Konge, P. Cayé-Thomasen, M.S. Sørensen, and S.A.W. Andersen, "Decentralized Virtual Reality Training of Mastoidectomy Improves Cadaver Dissection Performance: A Prospective, Controlled Cohort Study", *Otol. Neurotol.,* vol. 41, no. 4, pp. 476-481, 2020.
[http://dx.doi.org/10.1097/MAO.0000000000002541] [PMID: 32176132]

[48]   S.H. Lee, H.Y. Jung, S.J. Yun, B.M. Oh, and H.G. Seo, "Upper Extremity Rehabilitation Using Fully Immersive Virtual Reality Games With a Head Mount Display: A Feasibility Study", *PM R,* vol. 12, no. 3, pp. 257-262, 2020.
[http://dx.doi.org/10.1002/pmrj.12206] [PMID: 31218794]

[49]   R.M. Baños, M. Espinoza, A. García-Palacios, J.M. Cervera, G. Esquerdo, E. Barrajón, and C. Botella, "A positive psychological intervention using virtual reality for patients with advanced cancer in a hospital setting: a pilot study to assess feasibility", *Support. Care Cancer,* vol. 21, no. 1, pp. 263-270, 2013.
[http://dx.doi.org/10.1007/s00520-012-1520-x] [PMID: 22688373]

[50]   P. Athilingam, M.A. Labrador, E.F.J. Remo, L. Mack, A.B. San Juan, and A.F. Elliott, "Features and usability assessment of a patient-centered mobile application (HeartMapp) for self-management of heart failure", *Appl. Nurs. Res.,* vol. 32, pp. 156-163, 2016.
[http://dx.doi.org/10.1016/j.apnr.2016.07.001] [PMID: 27969021]

[51]   S. Sarker, L. Jamal, S.F. Ahmed, and N. Irtisam, "Robotics and artificial intelligence in healthcare during COVID-19 pandemic: A systematic review", *Robot. Auton. Syst.,* vol. 146, p. 103902, 2021.
[http://dx.doi.org/10.1016/j.robot.2021.103902] [PMID: 34629751]

[52]   S. Porkodi, and D. Kesavaraja, "Healthcare Robots Enabled with IoT and Artificial Intelligence for Elderly Patients", In: *AI and IoT-Based Intelligent Automation in Robotics.* Wiley, 2021, pp. 87-108.
[http://dx.doi.org/10.1002/9781119711230.ch6]

[53]   P.H. Kahn, B. Friedman, and J. Hagman, "I care about him as a pal,", In: *AI and IoT-Based Intelligent Automation in Robotics.* Wiley, 2002, pp. 87-108.
[http://dx.doi.org/10.1145/506443.506519]

[54]   L.A. Durán-Vega, P.C. Santana-Mancilla, R. Buenrostro-Mariscal, J. Contreras-Castillo, L.E. Anido-Rifón, M.A. García-Ruiz, O.A. Montesinos-López, and F. Estrada-González, "An IoT System for Remote Health Monitoring in Elderly Adults through a Wearable Device and Mobile Application", *Geriatrics (Basel),* vol. 4, no. 2, p. 34, 2019.
[http://dx.doi.org/10.3390/geriatrics4020034] [PMID: 31067819]

[55]   N. Robert, "How artificial intelligence is changing nursing", *Nurs. Manage.,* vol. 50, no. 9, pp. 30-39, 2019.
[http://dx.doi.org/10.1097/01.NUMA.0000578988.56622.21] [PMID: 31425440]

[56]   "The Past, Present, and Future of Technology in healthcare", *Brightscout.* Available at: https://www.brightscout.com/the-past-present-and-future-of-technology-in-healthcare/(2018).

[57]   S. Kr.Sharma, R. K Modanval, N. Gayathri, and C. Ramesh, "Impact of application of big data on cryptocurrency", In: *Cryptocurrencies and Blockchain Technology Applications.* Wiley, 2020, pp. 181-195.
[http://dx.doi.org/10.1002/9781119621201.ch10]

[58]   N. Noury, "AILISA: experimental platforms to evaluate remote care and assistive technologies in

gerontology", *Proceedings of 7ʰ International Workshop on Enterprise networking and Computing in Healthcare Industry,* pp. 67-72, 2005.
[http://dx.doi.org/10.1109/HEALTH.2005.1500395]

[59]   S. U. O. Andrew, W. Roberts, and M. A. R. Laura Blakeslee, *The Population 65 Years and Older in the United States.* US Department of Commerce, Economics and Statistics Administration: US Census Bureau, 2018.https://taubfoundation.org/files/2019/11/ACS-38.pdf

[60]   T. Barnay, and S. Juin, "Does home care for dependent elderly people improve their mental health?", *J. Health Econ.,* vol. 45, pp. 149-160, 2016.
[http://dx.doi.org/10.1016/j.jhealeco.2015.10.008] [PMID: 26608113]

[61]   E. Demir, E. Köseoğlu, R. Sokullu, and B. Şeker, "Smart Home Assistant for Ambient Assisted Living of Elderly People with Dementia", *Procedia Comput. Sci.,* vol. 113, pp. 609-614, 2017.
[http://dx.doi.org/10.1016/j.procs.2017.08.302]

[62]   J. García-Alonso, C. Fonseca, Ed., *Gerontechnology: First International Workshop, IWoG 2018* vol. 1016. Springer International Publishing: Cham, 2019.
[http://dx.doi.org/10.1007/978-3-030-16028-9]

[63]   S.E. Dilsizian, and E.L. Siegel, "Artificial intelligence in medicine and cardiac imaging: harnessing big data and advanced computing to provide personalized medical diagnosis and treatment", *Curr. Cardiol. Rep.,* vol. 16, no. 1, p. 441, 2014.
[http://dx.doi.org/10.1007/s11886-013-0441-8] [PMID: 24338557]

[64]   S. Armstrong, "The apps attempting to transfer NHS 111 online", *BMJ,* vol. 360, no. Jan, p. k156, 2018.
[http://dx.doi.org/10.1136/bmj.k156] [PMID: 29335297]

[65]   C.E. Kahn Jr, "From Images to Actions: Opportunities for Artificial Intelligence in Radiology", *Radiology,* vol. 285, no. 3, pp. 719-720, 2017.
[http://dx.doi.org/10.1148/radiol.2017171734] [PMID: 29155645]

[66]   L.D. Jones, D. Golan, S.A. Hanna, and M. Ramachandran, "Artificial intelligence, machine learning and the evolution of healthcare", *Bone Joint Res.,* vol. 7, no. 3, pp. 223-225, 2018.
[http://dx.doi.org/10.1302/2046-3758.73.BJR-2017-0147.R1] [PMID: 29922439]

[67]   P. V. William, D. Eggers, and David Schatsky, "AI-augmented government: Using cognitive technologies to redesign public sector work", *deloitte insights.* Available at: https://www2. deloitte.com/us/en/insights/focus/cognitive-technologies/artificial-intelligence-government.html (2017).

[68]   T.Q. Sun, and R. Medaglia, "Mapping the challenges of Artificial Intelligence in the public sector: Evidence from public healthcare", *Gov. Inf. Q.,* vol. 36, no. 2, pp. 368-383, 2019.
[http://dx.doi.org/10.1016/j.giq.2018.09.008]

[69]   A. Akmal, R. Greatbanks, and J. Foote, "Lean thinking in healthcare – Findings from a systematic literature network and bibliometric analysis", *Health Policy,* vol. 124, no. 6, pp. 615-627, 2020.
[http://dx.doi.org/10.1016/j.healthpol.2020.04.008] [PMID: 32456781]

[70]   S. Reddy, S. Allan, S. Coghlan, and P. Cooper, "A governance model for the application of AI in health care", *J. Am. Med. Inform. Assoc.,* vol. 27, no. 3, pp. 491-497, 2020.
[http://dx.doi.org/10.1093/jamia/ocz192] [PMID: 31682262]

[71]   G. Gopal, C. Suter-Crazzolara, L. Toldo, and W. Eberhardt, "Digital transformation in healthcare – architectures of present and future information technologies", *Clinical Chemistry and Laboratory Medicine (CCLM),* vol. 57, no. 3, pp. 328-335, 2019.
[http://dx.doi.org/10.1515/cclm-2018-0658] [PMID: 30530878]

[72]   A. Garg., "CoReS-Respiratory Strength Predicting Framework Using Non-invasive Technology for Remote Monitoring During Heath Disasters", In: *Book Chapter* AAP: US, 2021.

CHAPTER 4

Perspectives on Augmented and Virtual Reality (AVR) in Education: Current Technologies and the Potential for Education

S. Christina Sheela[1,*], **V. Selvalakshmi**[2] **and S.P.S. Arul Doss**[1]

[1] *Gnanam School Of Business , Sengipatti, Tamil Nadu 613402, India*

[2] *Srm Valliammai Engineering College , Kattankulathur, Tamil Nadu 603203, India*

Abstract: By combining information in the form of image alternatives with a software programme that stores knowledge on real images, augmented and virtual reality (AVR) technologies aid in the explanation of concepts. This methodology is developed to improve educational learning through two-dimensional media in education. In bloom taxonomy approach to teaching, integrating AR technology with academic content results in a new kind of automated application that serves to reinforce the usefulness of teaching-learning for learners in real-world situations. The learning outcome, which includes knowledge level and performance engagement, has a significant impact on all phases of higher education, from course planning to student evaluation and grading. AVR is a novel technique that combines elements of omnipresent computing, tangible computing, and social computing. This mode offers different affordances, combining the physical and virtual worlds, with continuous and implicit purposes of reading and interactivity. Digital resources are high-potential educational technology that enhances learning by supporting the learning environment through numerous e-resources. The various universities and Technical and Vocational Education Training (TVET) institutions give students an opportunity to complete an experiential learning component in their studies in order to complete their qualifications with the help of AVR implementation. This chapter provides an introduction to Augmented and Virtual reality (AVR) technology, the current status in education from different viewpoints, key technologies, and strategies mentioned in the context of higher educational learning output of students through these applications.

Keywords: Augmented Reality, Technology Augmented Reality in Higher Education.

---

* **Corresponding author S. Christina Sheela:** Gnanam School of Business, Sengipatti, Tamil Nadu 613402, India; Tel: 95858-81102; E-mail: christina.sheela@gsb.co.in

Adarsh Garg, Valentina Emilia Balas, Rudra Pratap Ojha & Pramod Kumar Srivastava (Eds.)

INTRODUCTION

Due to the recent rapid advancements in technology, Augmented Reality (AR) has emerged as a sophisticated technology for communication and knowledge exchange [1-4]. Azuma (1997) described augmented reality (AR) as having the following three features: (a) the integration of real and virtual objectives; (b) the interactive and real-time display of images; and (c) the registration (aligning) of augmented reality, and virtual items with one another. Two Boeing airplane engineers named Thoman and David founded AR in 1992. The first experiment's objective is to build a display that can alter the physical world. With this, virtual reality is taking the place of the real world. It requires more sophisticated mechanical arms and programmatically coded machinery to show [5]. It has industrial, medical, and military uses. AR is being used in a variety of fields, including entertainment, tourism, architecture, and marketing [3, 6, 7].

Augmented and Virtual reality (AVR) begins with the most powerful internet itself. It enhanced effective ways for learning and technology to teach complex knowledge. This technology is a theoretical and pictorial representation but left the part of real-time factual reality. Augmented Reality (AR) was designed with three-dimensional digital elements in a real environment to alleviate these drawbacks. Students look at these facts from several perspectives. 360-degree computer-generated images are used in virtual reality (VR) to recreate the virtual world in front of users wearing headsets. Through the application of this technology, people are given the idea that they are physically moving through and engaging with virtual environments. The primary distinction between these technologies is that whereas AR is used to create digital items within the context of the real world, VR is used to engage pupils in a virtual environment. India's higher education system is about to hit a fascinating inflection point. Digital technology is already being used increasingly frequently in educational settings. Higher education is currently being pushed into the experiential sphere by the forces of augmented reality (AR) and virtual reality (VR).

RECENT DEVELOPMENTS AND HISTORY OF AR

It has been thought that AR would improve the knowledge, perceptions, and communication of students. These technologies have the power to boost productivity in global tasks. However, until the 1990s, inertia became important. AR books are important stepping stones that connect the physical and digital worlds for the general audience. The use of AR technology in the classroom can benefit learners with three-dimensional (3D) shows and asynchronous experiences that are attractive to learners (Fig. **1**).

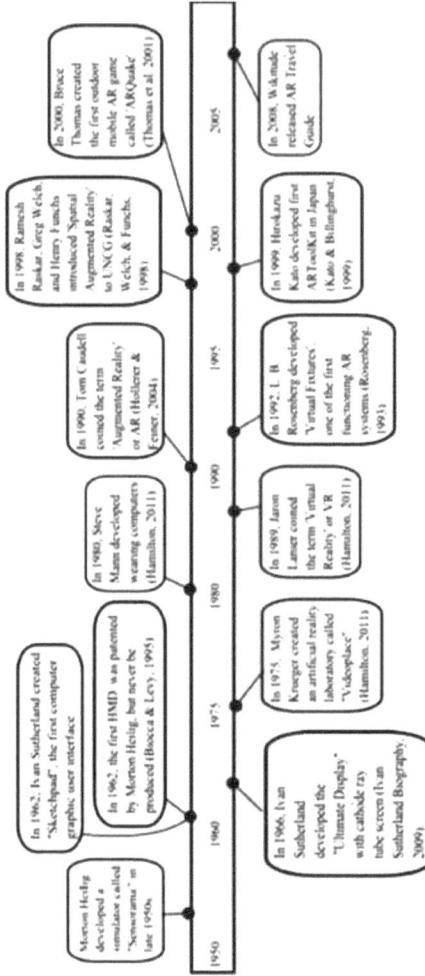

Fig. (1). History of Augmented Reality.

WHAT CHANGES CAN AUGMENTED REALITY MAKE TO THE EDUCATIONAL PROCESS?

AR has its own magic, as I referred to earlier. AR is a platform of technology wherein learning in the classroom becomes more transitional and active. The basis of augmented reality is that it combines many computer-generated graphics with the real world on screen. This indicates that AR enables you to view a computer-generated object on your display if you move your mobile camera in real life. Everything really transpires as you watch it on your webcam in real-time. The technology of augmented reality enables the direct transmission of 3D dimensions to insert, synthesize, and overlay digital and virtual information in the physical world. Only with the help of a smartphone, AR may be used to rapidly obtain information, visualize objects, and recognise objects. With the aid of this method, children may be able to learn in a more engaging setting. The AR experience also incorporates 25% digital reality and 75% of the real world. This means that AR incorporates virtual things into the real world rather than replacing the entire environment with a virtual one. By including videos, audios, or images in a real-world setting, you can experience an audio-guided tour [8].

The real goal of augmented reality and virtual reality in higher education is to increase student engagement and foster better comprehension. Using VR headsets to make learning engaging and interactive, instructors may convey difficult concepts to students in a controlled setting. AVR concentrates on the following three key traits: alignment of the 3D model to embed in the focused area; integration with a real-world environment; and real-time integration.

AR PERFORMS WITH TECHNOLOGY

The augmented reality applications were created and designed to be compatible with mobile hardware. This application was created using Unity3D and Vuforia, two different tools. The Vuforia framework enables the creation of augmented reality applications for mobile devices that use targets or patterns in the form of photos or objects. The production of augmented reality applications using Unity Technology is made possible by the integration of the cross-platform game engine Unity3D with Vuforia. With the help of Unity3D's features, it was possible to link both dynamic and static content to the targets and (ii) develop a user interface with toggle buttons that let students engage with the content by displaying words, replaying instructional audio, and presenting 3D visualisations. For instance, the learner can point to the structure of an atom using a set of toggle buttons, and the programme will display and explain the structure while modifying the content to be static or dynamic and displaying audio and text. The mobile device's augmented reality application can be seen in Fig. (**2**).

- With more enterprise adoption, in 2021, AR is increasing rapidly more than virtual reality. The AR market will grow by 27% CAGR from 2019 to 2026 that reaches USD 3664 million from USD 849 million in 2019.
- According to the report, AR in retail will grow at a CAGR of 20% in healthcare, at a CAGR of 27% in automotive, at a CAGR of 10% in the market, for the head-mounted and head-up projections, it is at a CAGR of 22% and 17%, respectively, and mixed reality AR at a CAGR of 68%.

- The use of AR and VR in the industry is being driven by their acceptance in the healthcare, education, remote meetings/conferences, gaming, entertainment, and e-commerce sectors. Fortune 500 companies have begun experimenting with these technologies. With the usage of cell phones and laptops, driving adoption skyrocketed.

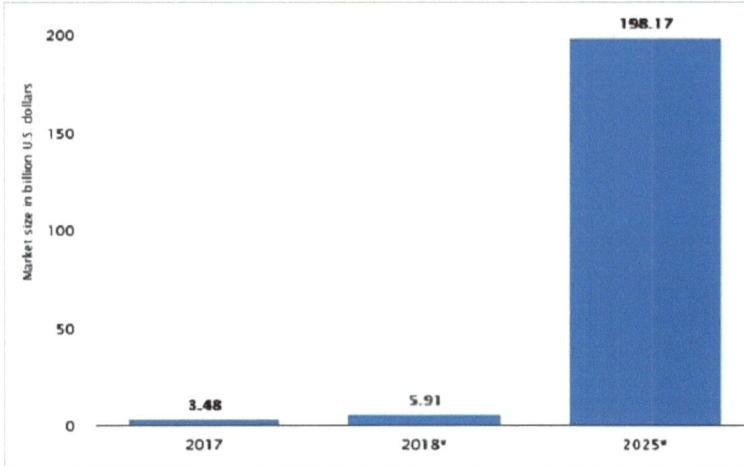

Fig. (2). Market share of AR.

AUGMENTED REALITY AND ITS ELEMENTS

The following components should be present in an AR system in order to achieve this superposition of virtual elements in the perspective of the physical environment: Fig. (**3**).

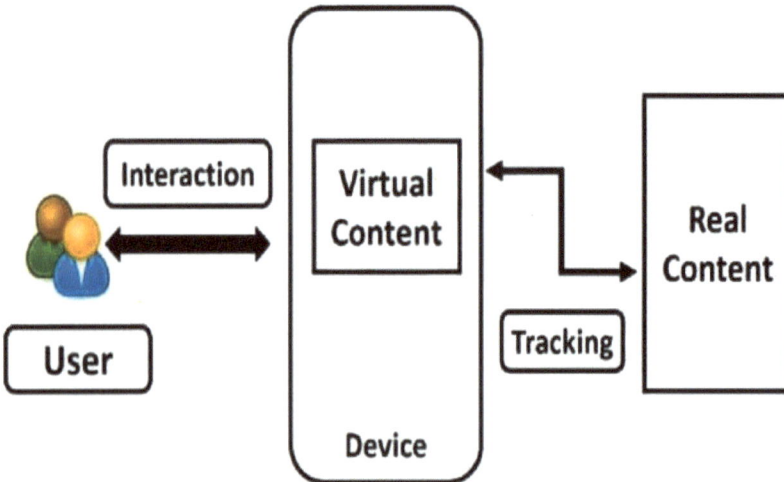

Fig. (3). Elements of AR.

Camera: The device that captures an image of the real world can be a smartphone camera, a tablet camera, or a webcam.

Processor: This hardware component combines an image with additional information that has to be superimposed.

Software: A special programme oversees the operation.

Screen: Display of a real-world and digital image.

Internet Connection: Using an internet connection, you can transfer information from a physical environment to a distant server and retrieve related virtual data that overlaps.

Activator: An activator is something that the device sends from a real-world element to the programme to allow it to detect the physical surroundings and choose the appropriate virtual information to be added. Examples include a QR code, bookmark, image, or item.

Marker: An element that is more common in 3D augmented reality systems is the marker. They can be made of paper or real-world objects, but they all have distinctive patterns that are simple for cameras to pick up and process and are visually independent of their surroundings.

OPERATION OF AR SYSTEMS

With the help of GPS, there are applications available to manage enhanced liveliness. The programme uses computer vision technology to examine the video stream and identify objects when a user points the gadget at it and stares at it. There are four varieties of augmented reality:

1. Marker-free Augmented Reality, AR references to a product application that holds a virtual 3D object to a fixed location in space without the necessity for prior knowledge of a client's current situation.

2. **Mark-based Augmented Reality,** AR application appraises the position of the camera in the casing and the direction of the genuine world.

3. **Projection-based Augmented Reality,** AR is also called spatial AR, a technique for conveying advanced data inside a fixed setting. Target clients and items can move around in the area; yet, the territory wherein AR is used is restricted to the fields taking into account the fixed projector and the supporting camera.

4. Superimposition-based Augmented Reality is the type of AR that relies on superimposition to replace things from the actual world. The virtual perspective of

an item is typically completely or partially replaced by an extended view in augmented reality.

Hardware

Augmented Reality hardware is head-mounted, wearable, or handheld. Some of the popular AR hardware include mobile phones and tablets; head-up displays (HUD); HMDs (Magic Leap); AR glasses (Vuzix Blade, Google, Apple, Facebook, Optinvent, Everysight Raptor, Epson Moverio, ThirdEye); and AR contact lenses or virtual retinal displays (VRD). A few types of AR glasses are shown in Fig. (**4**).

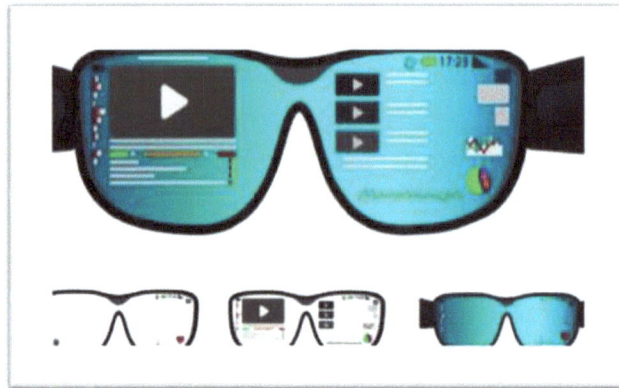

Fig. (4). Augmented Reality Glasses.

Software

AR applications can be developed easily with the recent software bundles available. The tools can be selected based on support for high-level abstraction, loading of 3D models, powerful and fast 3D rendering, support of different media formats, support for real-time tracking, a scripting language (C#, Javascript), open Scene Graph, interaction with IoT and IIoT, multiple platforms (Windows, Android, IoS), cloud-based targets and media, efficient computer vision-based image recognition techniques, easy camera calibration, single or multiple targets in a scene, and registration. Finally, the software can be selected based on open source or free, free for academic or commercial purposes. The popular software tools available in the market are Vuforia, Metaio, Unity3D with AR Foundation, Unity3D with Vuforia, Wikitude, ARCore, ARKit, MaxST, and Amazon Sumerian. Other tools may be required for the design of models and special hardware-specific libraries, in addition to these standard development tools of AR.

VIRTUAL REALITY

Introduction

VR is an emerging technology that creates a virtual environment for users to enable them to get an aesthetic feel for the desired surroundings and be fully immersed in the virtual world. VR technology is useful in advanced sectors such as engineering, medicine, education, entertainment, design, planning, construction, training, and entertainment, all with varied applications. VR tries to mimic the real world to give 3D immersive visual and interactive experiences by bringing depth to objects with an unlimited 3D space. VR immersion experiences show the ability to track user's interactions (telepresence: head, eyes, and hands) and it includes 3D simulated images that resemble full-sized pictures from the user's viewpoint correspondingly it regulates the images to reflect the changes on the user's screen.

Hardware

VR hardware is a kind of head-mounted closed display unit together with interaction devices that can be obtained with 3DOF (Oculus Go) or 6DOF (Oculus Quest, HTC Vive). The first VR headgear, Google Cardboard, was made possible because of the inertial measurement unit (IMU), which uses sensor fusion to monitor linear and angular acceleration and position changes. The availability of such high-performance GPUs makes the VR provide high-fidelity graphics. Some VR devices (HTC, Vive) work together with a powerful computer or mobile phone, and some of the recent VR devices (HoloLens, Oculus Go) have the computing power integrated. Microsoft Holo Lens, HoloLens, HTC Vive, and Google Cardboard alt-text Images of various VR hardware are used for visualizing and experiencing virtual reality. Fig. (**5**) shows a few VR devices used for various applications.

Fig. (5). Microsoft HoloLens, HTC Vive and Google Cardboard.

AUGMENTED AND VIRTUAL REALITY IN EDUCATION

Students can learn their lessons *via* images, videos, and animation by adopting AR and VR in education. These technologies have captured the attention of students across the country to achieve a higher rate of retention and increased knowledge in academic levels. Changing the place and method of learning material is the potential of AR, which can introduce new ways and methods. Attending the classes through AR will be more engaging and informative so that students can comprehend concepts. The spokespeople for education are aware of the fact that the learning process includes interaction and creativity.

Most learners have their own smartphones and they are active smartphone users. Modern students are using mobile devices for educational purposes, such as, doing homework assignments, accessing subject-oriented matters, *etc.* Student gets ample information easily through AR & VR. The amalgamation of both phone and AR, especially for educational purposes, is really popular, although still it should be exploited further to get the maximum benefit out of it. Now, AR satisfies the needs of learners related to subjects and their knowledge. Of late, the power to link reality and digital content is being enhanced gradually, and enough opportunities are provided for learners and teachers as different technologies are introduced (Fig. **6**). AR possesses enough power to be empowered as an innovative technology in the domain of education. Because of the monotonous style of learning, involvement in the classroom of students is getting reduced and the quantum of interest has been lost in modern days, but through technology, teachers can create a healthy learning environment. It encourages learners to obtain knowledge faster. In the current conventional scenario of education, teaching and learning have become boring. Teachers should know about how students will experience knowledge once they have learned it. Teachers have to update on this technology of AR in the classroom will create a system that teachers need to implement virtual reality in education.

Fig. (6). AR in Classroom.

Utilisation of AR & VR in Education

AR & VR classes help to motivate students, capture their attention as a whole, and help them understand difficult subjects. Following are some of the significant uses of AR in education:

Educational Itineraries

The educational value of itineraries for study tours, museums, exhibitions, *etc.* has been increased and made easier by AR. A new layer can significantly increase the amount of information displayed by animations, text, photos, videos, music, *etc.* These itineraries might be used by students as a model for their research projects.

3D Models

For learning with 3D models and graph paper, there are ideas and processes for measuring the area of a square, which is most approporate.

Simulations

In the initial stage, AR in simulation enables intriguing aspects like ease of use, accessibility, immediacy, practicability, safety, trial-and-error learning, cost reduction, *etc.* A fire drill is an illustration of a simulation. In this scenario, simulators using AR technology are used to teach individuals how to safely handle situations like fire, fire-fighting, coated electrode welding, or preparing a steak.

Real-Time Programmers

In the actual world, the text that already exists can be converted into a picture using AR technology. Word Lens is a smartphone and tablet application that uses the camera to take an image, detect the text in that image, and then translate it into the desired language before displaying the image on the screen and replacing the original text with the translation.

Accessibility

Access to public spaces and resources can be improved with the use of AR. Due to the increased likelihood of access to information, this is particularly beneficial for kids who have educational requirements. The Research Project serves as an illustration of AR for accessibility.

Benefits of Using AR in Education

The development of the AR application in education is revised by the educational division. This paradigm shift has been completed, and it continues to magnify in all fields of the educational domain. Thus, the advantage of AR in education is totally different owing to its remarkable impact. It is understood that AR will still process education and the learning method, as below.

1. Increasing students' interest and classroom engagement

2. Practical learning

3. Cheaper cost of learning

4. 24×7 available study materials

5. Growth of memory

Nurturing Training Method

The utilization of AR in the academic domain creates edutainment and entertainment for college students. The reality is that the students instigate higher learning methods through AR. It attracts students to the learning track wherever it becomes important to analyze new concepts, and it will increase the inventive art of scholars within the learning method. Finally, students develop ingenuity and the ability to discover and acquire additional concepts through this training method.

Increasing Students' Participation in Classes

AR will increase the participation of a student in the classroom. With the help of AR apps, pupils possess the chance of using coaching training models. These models facilitate the high-level acquisition of concepts, understanding, and quick delivery among students. Finally, students will have a high level of interest, leading to dynamic schoolroom participation, and increasing the motivation and interest of scholars within the learning method.

AR Increasing Memory Capacity

AR improves the educational interest and enthusiasm among learners and students, and it will improve imagination and increase the memory power as well.

Interactive Lessons by AR

Using AR in the education domain offers asynchronous lessons to pupils through a clear and higher level acquisition of subject ideas and conjointly the simplest ways to interact with students. It increases informative activities and also enhances learning of students.

Increasing Sensory-Motor Development

AR offers interactive privileges to better students' physical and psychological faculties. This leads to improved sensory development and facilitates an academic learning method to an entirely innovative stage.

Less Expensive

The price of subject materials, 3D models, posters, and the prototypes is typically high. The academic centers do not have the resources to purchase and maintain these materials. These learning models need to be available to pupils in classrooms at the same time in their homes as well. AR minimizes the cost of buying learning apparatus and is more cost-effective for a long period as users no longer have to spend money on physical materials repetitively.

Enriched Ways of Telling a Story

The use of AR in education promotes the scripting of academic classes *via* optical models, which, further facilitates the students to improve academic ideas and the concept of development.

Increasing Learning Activities

Technology has emerged as a compulsory element of education, and students fully rely upon technology. Interactive models provide access to enhanced learning activities and interactive activities for college students.

Visiting the Past, Present, and Future

AR in education is decisive and encourages students to induce knowledge of the past, present, and future. Therefore, an opportunity is given to practice data to undo issues and visit the past, present, and future. This further increases the enthusiasm and observation within the learning method, thus playing a significant role.

AR IN LEARNING AND EDUCATIONAL DOMAIN

Five Directions of AR in Educational Environments

Discovery-Based Learning

AR can be applied in discovery-based learning, along with concepts pertaining to real-world situations like galleries, planetariums, and ancient places.

Objects Modelling

AR may be employed in modelling applications. These types of object modelling develop interest among learners to obtain immediate visual feedback on what a particular element will look like in different settings. These applications also allow students to engineer virtual objects by analyzing the physical features or communication among objects; like architectural education.

Augmented Reality Books

The importance of AR books increases with the assistance of technological devices like special glasses, enabling the students to use 3D presentations. At the first level of learning, AR books will be an instructional medium.

AR Gaming

AR games are primarily helping learners to understand ideas. It is constructed within the real world, and increased networked information will provide effective ways to point out relations and linkages. These AR games are applied in anthropology, history, social science, earth science, *etc.* which make 3D settings viewed through a mobile device or camera.

Skill Training

AR will reflect abilities in an instructive capacity. AR has a strong foundation for amazing objective learning in situational learning experiences and unforeseen examination, however, simultaneously, emphasizing how abstract concepts are interconnected in reality. AR excels at providing contextual learning opportunities, extensive analysis, and further investigation.

Multidisciplinary Use of AR

AR is used in several independent disciplines like medicine, design, diversion, tourism, gaming, and education. Using AR in environmental education increases confidence and encourages self-paced learning. In higher stages, AR shows enhancements in previous methods in terms of engrossment and inspiration [9]. It says that the use of AR is appropriate in every academic system. It is considered that in academic systems, AR is indeed a real tool that will accelerate the learning pace to obtain better learning outcomes through pleasant ways and means, and a convenient method for both teachers and students at large.

Place of AR in Classroom Learning

In teaching and learning platforms, AR as technology provides interactive digital features text, pictures, videos, sounds, 3D objects, and computer graphics – in real-world environments. A variety of such applications promote skills like originality, problem-solving, crucial thinking, analysis, coding, and pragmatic testing through online platforms. Many teachers have made an effort to write their own course plans. AR in education especially encourages students to study facts on the subjects and the inherent reality of the content. Scanning a book's outline might provide students with a brief outline of the book. There are certain applications that are already available that enable students to scan book covers for reviews. In modern times, the procedures used in the teaching and learning process in the classroom are listed below.

AR-Enabled Worksheets

Worksheets help the students to use AR technology from their own place. The monotonous process of homework can be converted into a simple one for students, and they can explore both AR and content in their homes. These worksheets will help students develop a positive relationship between technology and education.

Faculty Photo Walls

The faculty photo wall has pictures of faculty members on a display board which is a creative method to integrate AR into education. Using AR, students can scan images for their future reference. This is out-of the-box thinking, which enables learning more information about teachers and getting all possible information. An AR-based word wall in a collaborative environment allows learners to see the meanings and draw the key vocabulary words as a whole. In this way, students use their mobile devices to learn the use of words and sentences in order to improve their vocabulary. AR can also be used for parent involvement by tracking their regular activities being held at institutions.

Custom-Made Markers

Marker images serve the purpose of displaying AR content over a surface and directly linked to the lesson content. Students take images in a camera with AR textbook pages, and the AR application identifies it and sends a query to the server immediately. The camera in the mobile device pointed at the page primarily serves as a registering device for the orientation and movement of an image that allows an object to rotate at all angles. It is implemented today in a widespread of applications in education. Faculty can use Aurasma and Layar, which are two application online programs, in the form of video, internet content, or a non-dynamic image.

Premade Resources

It is a good and easiest platform to start for teachers with AR technology in classrooms. Pramade Resource content helps teachers take away the stress of getting involved in the technological phase of lesson planning and to carry on with the subject. It is a very excellent inbuilt process for all learners and they require a smartphone to bring the textbook to reality. Carlton Books is a leading international publisher that publishes beautifully illustrated AR books for both adults and children.

Augmented Reality in the Classroom

There are five different ways to use AR in the classroom:

1. Aurasma: They are provided with online content in a viewing window that proves to be highly engaging for students.

2. Daqri Studio: It gives a thriving realm of narrative knowledge transmission and 4D encounters through AR.

3. Quiver: The AR coloring app that creates engaging, immersive coloring experiences for people of all ages.

4. Fetch Lunch Rush: AR multi-player game.

5. Aug: 3D and 360-degree views of the animated lessons come in a variety of formats.

Using AR Education App for Learning and Development

In the domain of education, the following AR tools are used for learning enhancement:

- 3DBear
- Catchy Words AR
- Co Spaces Edu
- Froggipedia
- Jig Space
- MERGE Cube
- Metaverse
- Boat boat
- Orb
- World Brush

Augmented Reality Applications for Classroom

Aurasma: Students will utilize this app to convert their learning to real life.

Preparation Mini-Lessons: Students scan a page of their work to display a video of the teacher attempting to solve the challenge during the preparation of mini-lessons.

School Picture Wall: This app helps students to understand the environment through photos and images.

Book reviews: Students can add some "air" to a book by recording themselves, delivering a brief evaluation of something wonderful they've just finished. After that, anyone can check the book's foundation and immediately access the survey.

Yearbooks: The ways in which augmented reality will enhance a yearbook are countless, ranging from recognition for video profiles to sports features to productions and show films. **Lab Safety:** Place triggers (pictures that enact media once checked by a gadget) all around a working environment where students will learn different well-being strategies and conventions for research facility instrumentality.

Difficult of Hearing (dhh) and Hard of Hearing Language Cheat Sheets: With AR, jargon word cheat sheets for the deaf will include a video overlay that explains how to use a phrase or expression.

World Simulations with Interactive Objects: This app will create a lot of work by adding clickable objects and hot spots which create simulations on the content.

Immersive Stories: AR stories are one of the foremost emotion-oriented devices. AR books create new readers' fictional environments, and e-Learning personnel will weave stories that draw online learners in and create comfort with e-learning characters.

AR Resource Links that Feature Stats and Facts: Any individual will scan a unique AR code or trigger the object to be told questions. For example, a science textbook would possibly take readers to a virtual laboratory wherever they do experiments. AR materials linked to e-Learning materials, and these will disclose fascinating facts, statistics, and concepts that enable them to explore on their own.

Three-Dimensional Learning Models: It is the most advanced use of AR in e-Learning with 3D and demos. Online learners would read through pictures, figures, and maps on their monitors. For instance, a clickable 3D diagram of producing instrumentality throws light on key options and the functions of every half will be flexible to govern these 3D visualizations.

On-line Group Collaboration Projects: e-Learning AR is not only for interactive usage but for sure it permits online students from around the world to move with their fellow learners in virtual settings. It permits two efficient uses in online coaching. Workers can attend a live company e-Learning incident such that options increased live activities in live social settings. Another use is online group collaboration.

IMPLEMENTATION OF AR DETAILED IN NATIONAL EDUCATION POLICY 2019

Technology Use and Integration in Instructional Settings

The New Education Policy 2019 (draft) conjointly emphasized the importance of ICT, AR, and VR in education and recommended to implement an equivalent teaching-learning method for approaching the education level. Technology implementation and use are going to be pursued as a very significant objective for raising the quality of education. Therefore, the main objective is not only delivering prime excellence content, however, conjointly, using suitable technology to support the translation of content into manifold languages, help learners with specific needs, and improve the teaching–learning pedagogy standards and processes through the use of smart tutoring systems and adaptive assessment systems. The use of technology will generate new immersed content to strengthen the instructional design and interactive management. It also brings clarity and potency to the assessment method as well as to the authority and corporate body in the process to assist with the management of education, like increasing the open and distance learning system and subsidiary teacher growth programs; so it will respond to the rising demand for faculty education for all ages, professions and courses of study, vocational training, and distance learning.

Training of Faculty Members

As educators become ready to implement AR, there are challenges to be faced with it. A number of these problems are general, and educators won't be facing them alone. Yet, it's vital for educators to remember them and present them to directors, technologists, and potential sources for rectification and learning (Fig. 7).

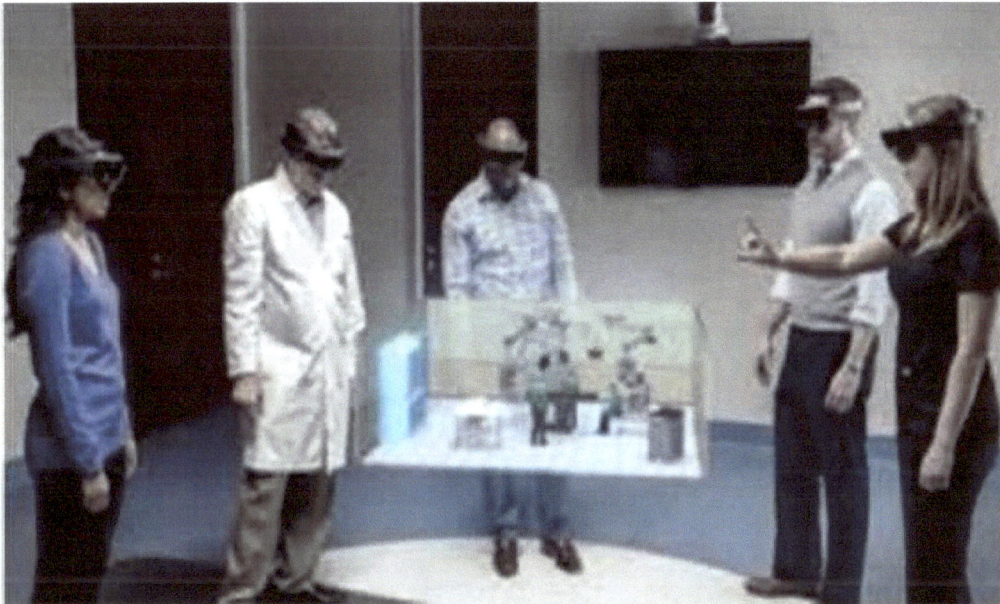

Fig. (7). Augmented Reality in Education and Training.

Planning and Administration for Implementation

AR is essentially within the classroom; the nominal AR setup might include:

- Web Affiliation
- Mobile Devices
- AR Apps
- "Triggers" or "Markers"

COORDINATED UTILIZATION OF AR & VR APPLICATIONS

Improvements in Augmented and Virtual Reality innovations have taken large steps in e-learning by:

1. Consistent Preparation

AR and VR made the students learn by rehearsing in a vivid virtual climate that can recreate this present reality climate.

2. Hands-on Preparation

This is an expert technique for preparing students with the ability, knowledge, and skills to execute a reasonable job inside their work environment.

3. Basic Abilities Preparation

AR & VR prepares companies to take advantage of specialists with basic abilities. For example, a car manufacturing company, airplane producer Boeing utilizes VR innovation to prepare their laborers. The molded VR pictures of the body of the airplane of Boeing airplane, with intensive, precise visuals of the wires that must be connected are shown in Fig. (**8**).

Fig. (8). Integrated Application of AR.

PREPARING IN UNSAFE CIRCUMSTANCES

Utility companies like power plants, oil and flammable gas, and substance producing businesses prepare representatives for their hardware and cycles, as well as for working in dangerous spots like mines, which is risky without satisfactory skills and information. The virtual climate created utilizing VR and

AR innovation assists the company with preparing students for testing and working in perilous circumstances. The computer-generated simulation can reproduce the genuine working challenges to prepare the laborers without the dread of committing any errors. Likewise, the student can rehash the learning interaction "n" a number of times to acquire dominance over it.

Representative Onboarding

Studies have shown that representative onboarding has an immediate connection with work fulfillment and workers' well-being. The representatives are destined to leave the company if there is no established orientation program. Companies are investigating approaches to take advantage of VR and AR in worker onboarding to make the experience more exciting and draw in fresh recruits. Likewise, VR can help recently added team members feel sure about their jobs.

Item Preparation

AR and VR can disturb current preparation strategies like study halls or conventional types of e-learning. To prepare laborers for item preparation, the company can incorporate AR and VR innovation and train specialists in a reenacted working climate.

Deal Preparation

AR and VR support the company with deal preparation capabilities, and item information and improve the mechanical development that makes it certain to track down an unmistakable spot in deal preparation.

Delicate Abilities Preparation

Utilizing AR and VR for delicate ability preparation can assist students with taking an individual learning path and making it more relatable. Subsequently, the company can long before see the more extensive utilization of AR and VR in figuring out how to prepare the specialists on delicate subjects. There are now numerous portable applications and assets accessible in the market today that utilize AR and VR innovation to prepare students for different delicate expertise prerequisites like Speech Center, VR, Virtual Orator, and Public Speaking Simulator VR.

Different Regions

Micro-learning technique is another inventive method bring students under e-learning through AR and VR.

Practical Online Assessments

With VR practice, the institution can plan the functional online evaluation patterns as miniature chunks for the students.

VR info Graphics

For making info graphics that convey a vivid encounter with the client, VR info graphics are the better approach.

Virtual Case Considerations

The depth learning experience utilizing AR and VR can assist with creating visual contextual investigations that can remake the situation. The utilization of vivid visuals makes the student understand a content-based contextual analysis.

CONCLUSION

AR is employed in almost every field. AR in education continues to be valuable. AR is an effective instrument for learning, creating the need for the knowledgeable role of individual learning by providing sharpness and interactivity within the instructional method. With the application of augmented reality (AR) in education, an effective interactive teaching-learning approach and this dynamic function of education are predictable. This is the manufacturing of a technologically savvy generation of youths. Although the possibilities of teaching in AR are wonderful and establish new ways of learning, academics ought to catch the eye of scholars and encourage them to obtain new tools to ascertain their subjects and complicated ideas and likewise to getting sensible skills. Moreover, even experienced adults can be benefited from taking part with youngsters in this review. AR in education is determining and encourages scholars to induce information on the past, present, and future. This is often a precise chance offered by utilizing AR in education and enables learners to achieve data of all three events. Therefore, a chance is given through AR & VR to practice data to undo issues and visit the past, present, and future. Therefore, increased enthusiasm and observation within the learning method are very significant. This can help learners, students, teachers, and employees to learn and upgrade their knowledge at any time and at any place. AR and VR are useful to the students to help them increase their knowledge by associating theory with practical applications and with repetition based on the interest and pace of the learner. This technology makes the students evaluate themselves. Gamification helps students to learn with interest by immersion rather than just by bookish theory learning. It is easy to adapt and develop new applications with limited or no programming using predefined functions or packages. AR and VR help an academic organization to update and adapt their skill set to the latest future needs of society, industry requirements, and frequently without much investment as well as follow the standards norms of Education 5.0 to be ready to adapt to Industry 5.0.

REFERENCES

[1]   R.T. Azuma, "A survey of augmented reality", *Presence,* vol. 6, no. 4, pp. 355-385, 1997.
       [http://dx.doi.org/10.1162/pres.1997.6.4.355]

[2]   R. Azuma, Y. Baillot, R. Bchringer, S. Feiner, S. Julier, and B. MacIntyrc, "Recent advances in augmented reality", *Comp. Grap. Appl.,* vol. 21, no. 6, pp. 34-47, 2001.
       [http://dx.doi.org/10.1109/38.963459]

[3]   M. Billinghurst, and H. Kato, "Collaborative augmented reality.communications of acm augmented reality in education. new horizons for learning", Available From: http://www.newhorizons.org/strategies/Technology/billinghurst.htm (Accessed on July 20, 2010).
       [http://dx.doi.org/10.1145/514236.514265]

[4]   M. Billinghurist, "Augmented reality in education, teaching and learning strategies", Available From: http://www.newhorizons.org

[5]   K.H. Cheng, and C.C. Tsai, "Affordances of augmented reality in science learning suggestions for future research", *J. Sci. Educ. Technol..* vol. 5, 2013, pp. 55-64. Available From: https://www.scirp.org/
       [http://dx.doi.org/10.1007/s10956-012-9405-9]

[6]   S.P. Yadav, and S. Yadav, "Fusion of medical images in wavelet domain: A discrete mathematical model", In: *Solid. Eng.* vol. 14. Cooperative University of Colombia- UCC, 2018, no. 25, pp. 1-11.
       [http://dx.doi.org/10.16925/.v14i0.2236]

[7]   R. De Lorenzo, "Augmented reality and on-demand learning", *Mob. Lear.,* 2009. Available From: http://mobilelearner.wordpress.com/2009/10/17/augmented-reality-and-on-demand-learning (Accessed on July 22, 2010)

[8]   I. A. C. Giglioli, F. Pallavicini, E. Pedroli, S. Serino, and G. Riva, "Augmented reality: A brand new challenge for the assessment and treatment of psychological disorders", *Adv. Comp. Psychometrics,* pp. 1-12, 2015.
       [http://dx.doi.org/10.1155/2015/862942]

[9]   Y. Huang, H. Li, and R. Fong, "Using augmented reality in early art education: A case study in Hong Kong kindergarten", *Ear. Child. Devel. Care.,* vol. 186, no. 6, pp. 1-16, 2015.
       [http://dx.doi.org/10.1080/03004430.2015.1067888]

CHAPTER 5

A New Approach to Crime Scene Management: AR-VR Applications in Forensic Science

Vinny Sharma[1,*], **Rajeev Kumar**[1], **Kajol Bhati**[1], **Aditya Saini**[1] and **Shyam Narayan Singh**[1]

[1] *Forensic Science, School of Basic and Applied Sciences, Galgotias University , Greater Noida, Uttar Pradesh 203201, India*

Abstract: The crime scene, a place where a crime has been or is suspected to have occurred or where the evidence related to a crime was found, is a vital part of the investigation as it contains all the major information about a crime. A keen eye for the crime scene can determine the possible *Modus Operandi* of a crime and establish *Corpus Delicti* in a court of law. As per Indian law, we are allowed to visit a crime scene once only and if we want to visit again, we have to take permission but it is of no use. During a revisit to a crime scene, the chance or probability of finding any evidence is nearly 0%. Documenting a crime thus became a very crucial step. By using ordinary methods for documenting the crime scene we cannot give a visual or walk to the actual crime scene. It's just a physical view of the documents. It is, therefore, critical to visually capture the crime scene and any potential evidence to aid the investigation. The current demand is for Augmented Reality and Virtual Reality. Virtual Reality is a wholly virtual view of the scenario, whereas Augmented Reality reflects a real-world context. VR is a type of advanced user interface that comprises a real-time simulation of a real-world environment with which the user interacts through numerous sensory channels: sight, hearing, touch, smell, and taste. The 3D reconstruction and visualization of crime situations such as criminal assaults, traffic accidents, and homicides is a new method of criminal investigation that has the potential to improve efficacy. To produce an accurate and immersive virtual environment, modern 3D recording and processing methods, such as AR and VR, are used. Immersion in a virtual environment, on the other hand, allows for various points of view.

Keywords: Crime Scene, Reconstruction, Augmented Reality, Virtual Reality, 3D Capturing.

INTRODUCTION

"An act that is illegal or against the law, and is subject to judicial punishment, is a crime." It is considered to be an intentional act of breaking the law that was perp-

---

[*] **Corresponding author Vinny Sharma:** Forensic Science, School of Basic and Applied Sciences, Galgotias University ,Greater Noida, Uttar Pradesh 203201, India; Tel: 91-8447723691; E-mail: vinnysharmashah@gmail.com

Adarsh Garg, Valentina Emilia Balas, Rudra Pratap Ojha & Pramod Kumar Srivastava (Eds.)

etrated by someone without cause. Many legal experts have given their own interpretations of what constitutes a crime. Following are a few definitions from well-known jurists:

Crime is defined by William Blackstone as "an act committed or neglected in violation of public law prohibiting or requiring it" in his book Commentaries on the Laws of England.

"A crime is a violation of a right, considered in reference to the ill tendency of such violation as regards the community at large," according to Sergeant Stephen.

The Indian Penal Code of 1860 and the Code of Criminal Procedure of 1973 neither utilize nor define the term "crime." Instead of "crime," the word "offence" is used. It is defined as "offence is an act punishable under the Code" in Section 40 of the Indian Penal Code, 1860 [1].

COMPONENTS OF A CRIME

There are four components that make up a crime, and they are as follows:

- Human being.
- Men's rea or guilty intention.
- Actus reus or illegal act or omission.
- Injury to another human being.

Stages of a Crime

Every crime begins with the intention to commit it, followed by the preparation, the attempt, and finally, the execution of the intention. The following are the stages:

1. Intention: Having the intent to commit an offence is the first step in carrying it out. However, the law does not take into account an intention; a crime cannot be committed if there is only an intention to conduct it and no subsequent act. The fact that it is so challenging to establish a person's guilty thinking is the reason why the "suspect" is not being prosecuted at this time [1].

2. Preparation: Preparation is the second step in the commission of a crime. It implies setting up everything needed for the performance of the intended act. The crime cannot be committed by intention alone or by intention combined with preparation. The legislation does not now make preparation punishable [1].

Note: Rare Situations where Preparation Is Punishment

- *Preparation is generally not punished, but there are some rare circumstances when it is. Here are some examples:*
- *Section 122 of the Indian Penal Code (IPC) 1860 prohibits preparing to wage war against the Government; Section 126 of the IPC 1860 prohibits preparing to invade the territory of a power at peace with the Government of India;*
- *Section 399, IPC 1860, "Preparation for Dacoity";*
- *Sections 233-235, 255, and 257 pertain to the preparation for counterfeiting money or government stamps.*
- *Having fraudulent documentation, bogus weights, or measurements, or counterfeit currency. Sections 242, 243, 259, 266 and 474 all define possession of these items as a crime, and no possessor may claim that he is still in the planning stage [2].*

3. Attempt: Following preparation, an attempt is the first step in the actual committing of a crime. A try must have these three components:

- Guilty intent to commit an offence;
- Some action taken to contribute to the commission of the offence;
- The action must not constitute the entire offence.

Note: Attempt, The Indian Penal Code, 1860 - The Indian Penal Code includes four different ways to deal with attempt [3]:

- *The same punishment is set forth for both completed offences and attempted offences in the same provision. Sections 121, 124, 124-A, 125, 130, 131, 152, 153-A, 161, 162, 163, 165, 196, 198, 200, 241, 251, 385, 387, 389, 391, 394, 395, 397, 459, and 460 contain such provisions.*
- *The commission of certain crimes and attempts to do them have been dealt with separately, and the penalties for attempting to commit such crimes are different from those for crimes that have already been committed. Examples include the penalties for murder under Section 302 and the penalties for attempted murder under Section 307. Another illustration is that robbery is prohibited under Section 392 while robbery attempts are prohibited under Section 393.*
- *Under section 309, suicide attempts are prohibited.*
- *Section 511, which stipulates that the accused should be punished.with one-half of the longest term of imprisonment imposed for the offence, a prescribed fine, or both, applies to all other situations.*

4. Achievement Or Completion: The accomplishment or completion of an offence is the last step in its commission. If the accused successfully performs the crime, he will be found guilty of the full offence; however, if he fails to do so, he will only be found guilty of an attempt [4].

CRIMINAL LAW

Criminal law is the body of law that outlines all criminal offences, governs how offenders are apprehended, and establishes the appropriate penalties, punishments, and/or forms of treatment.

In terms of criminal law:

1. IPC (Indian Penal Code), 1860

2. IEA (Indian Evidence Act), 1872

3. Cr. PC (Code of Criminal Procedure), 1973

Criminal: The term criminal refers to a person who commits a crime, disobeys the law, or is otherwise accountable for an illegal act.

Crime Scene

• The Crime Scene is the location of the crime. Any location that offers tangible proof, which is unquestionably the crime scene's beating heart.

• It is where the majority of the information for the investigation is obtained at the outset.

• The crime scene, often known as the scene of the incident, is not territorially constrained. It is not constrained and can be applied in multiple locations.

• The location of the crime scene is determined by the type of crime that was committed.

• There are never any two identical crime scenes.

• Each crime scene includes not just the physical location but also people and objects.

• The crime scene needs to be adequately guarded. If the contents of the region are not also protected, protecting the place is useless.

• The physical space and the tangible items contained therein are typically simple to secure. The persons at the crime scene must be kept as carefully as any other piece of evidence, which is more challenging.

• The crime scene consists of all places that the participants passed through when they entered the scene with the intent to commit the crime, carried out the crime, and left the scene [1, 2].

Types of Crime Scene

There are three classifications on different bases:

1. On the basis of the physical location of the crime & evidence:

  a. Primary crime scene:
  b. Secondary crime scene:

2. On the basis of size (Evidence or Crime Scene):

  a. Microscopic
  b. Macroscopic

3. On the basis of Location:

  a. Outdoor
  b. Indoor
  c. Vehicular

Crime Scene Investigation: The three concepts of science, logic, and law all come together in crime scene investigation. The goal of a crime scene investigation is to gather information about what truly occurred at the crime site, which is undoubtedly impossible for us to imagine. But the physical proof, such as any biotic or abiotic component found at the murder site, speaks for itself. It aids in establishing a clear connection between physical evidence, suspect, and victim.

Components of Crime Scene:

1. Line of Approach: The route used by the criminal to reach the crime scene after committing the crime. As an illustration, use this road to get to point A (Crime Scene).

2. Point of Entry: The location where the offender would have arrived at the crime scene to begin committing the offence. For instance, the front entrance, back door, balcony, or even the house's roof (which is a scene of crime here).

3. Actual Scene: The precise location where the crime really took place, where the victim was discovered, or where the majority of our physical evidence may be

found. For instance, a dead person might be lying on a table in the dining room or on a bed in the bedroom.

4. Point of Exit: The location from which the criminal must have fled, fled, or made his exit after committing the crime.

5. Line of Retreat: The route used by the criminal to flee "far" from the crime site. For instance, the route taken by the perpetrator to get to his house.

The forensic science field of crime scene reconstruction, also known as crime scene reconstruction, provides "specific knowledge of the sequence of events leading up to the commission of a crime using physical evidence, scientific procedures, logical and inductive reasoning, and their relationships. "Reconstructing a crime scene "involves looking at the context of a scene and the physical evidence obtained there in order to establish what happened and in what order it happened," according to Gardner and Bevel.

INTRODUCTION TO AR (AUGMENTED REALITY) & VR (VIRTUAL REALITY)

The unpopular reality of taxpayers we see is the interactive realization of real-world where computer-generated perceptual knowledge is used to add things to the real world, sometimes involving multiple senses, including olfactory, haptic, somatosensory, and visual sense. Actual-time interaction and precise 3D registration of virtual and virtual items are made possible by the technology known as augmented reality (AR), which mixes the real and virtual worlds. Covered sensory information may be helpful or harmful. These experiences are so closely linked to the real world that they are considered to be completely rooted in it. The genuine unpopularity of taxpayers we see alters a person's impression of a real-world space, whereas virtual reality replaces the user's real-world real estate with a simulated image [5, 6].

Augmented reality (AR) is a technologically advanced real-world image created by the use of digital visual components, music, or other sensory stimuli. It is a growing trend among businesses dealing with mobile computers and marketing applications. One of the main objectives of augmented reality, among the extension of data collection and analysis, is to emphasize certain aspects of the real world, to raise awareness of those qualities, and to produce intelligent and accessible information that can be applied to real-world systems. Big data can help organizations make better decisions and gain an understanding of customer purchasing practices, among other things [7, 8].

A virtual experience called virtual reality (VR) can be both similar to and different from the real world. There are many applications for virtual reality in business, education, and entertainment. A computer-generated world with realistic-looking visuals and objects is known as virtual reality, and it gives viewers the impression that they are entirely involved in their surroundings. Through the use of a virtual reality helmet or headset, this world is seen. We may use virtual reality to immerse ourselves in video games as if we were one of the characters, learn how to conduct heart surgery, and increase the quality of sports training to boost performance. Everything we see is part of an artificially built world created through pictures, sounds, and other means [9 - 11].

AVAILABLE TECHNIQUES FOR SOC (SCENE OF CRIME)

As is common knowledge, a crime scene is a specific location where any illegal conduct has occurred. Therefore, there cannot be a single correct method for handling every crime scene. It is necessary to assess each scene separately.

Logic and viability are two key components of crime scene investigation. Processing a crime scene requires reasoning about what might have happened, and the viability of that occurrence is then examined. Once these two problems are rectified, a thorough search for any missing physical evidence takes place, followed by recognition. The gathered physical evidence was properly packaged at the scene and delivered to the forensic science lab along with a list of needs for laboratory-based scientific examinations. These scientific perspectives not only assist us in minimizing the need for eyewitnesses, but they also establish connections between the suspect and victim and the murder scene. The victim and the crime site are generally known or accessible in all cases of crime; the only thing missing is the suspect or criminal who actually committed the crime. In order to identify the suspect, physical evidence that is present with the victim, at the crime scene, or occasionally even with the suspect, must be used. The crime scene processing must go through a number of stages in order to accomplish the aforementioned goal [12].

Physical evidence analysis can be divided into two main phases. Crime scene investigation is the initial stage of the investigation process. It begins at the crime site, which serves as the meeting spot for those responsible for the crime and the location where traces are exchanged. The physical evidence from the crime scene needs to be identified, noted, gathered, and packaged. After that, the "Laboratory investigation" phase, which is the second stage, began. In this phase, all the evidence gathered from the crime scene is sent for scientific analysis in accordance with the requirements of the investigating officer, after which a report is created. Finally, the report is provided to the court so that it can make a decision on the case in question based on the findings of the scientific investigation [13].

Therefore, the entire procedure may be broken down into the following sequence of stages for conducting a good crime investigation:

1. Protection of crime scene

2. Recognition of evidence

3. Searching of evidence

- Spiral
- Grid
- Strip
- Quadrant or Zone patterns

4. Documentation of crime scene and evidences

- Report/Note making
- Sketching
- Photography
- Videography

5. Collection of Evidence

6. Marking of Evidence

7. Packaging of Evidence

8. Analysis of Evidence

9. Interpretation of results

10. Reporting of results and expert testimony

PROS AND CONS OF AVAILABLE TECHNIQUES FOR SOC

To document a crime scene, there are these four techniques that are generally used at a crime scene, these are:

  i. Report/Note Making
 ii. Sketching

iii. Photography

iv. Audio & Videography

Report/Note Making: Even when additional recording methods are used, written notes still serve as the cornerstone of the crime scene and physical evidence documentation techniques and are crucial to any investigation.

Sketching: The best way to depict the spatial relationships of the things present at crime scenes with all significant dimensions accurately captured is through sketches rather than notes or photographs, which are both critical and essential components of any crime scene documentation. Therefore, sketches are a crucial part of crime scene documentation. If a preliminary sketch of the crime scene isn't made there, it won't be possible to document the scene properly.

Photography: Regardless of how skilled an investigator is at describing a crime scene, pictures might be claimed to be able to tell the same tale more clearly and accurately. Everyone is aware of the value of photos in crime scene analysis. It is the most straightforward and effective method of recording a crime scene, together with criminal sketches. Photographs are a very helpful tool for presenting minor but important pieces of evidence with accurate measurements of the crime scene.

Audio & Videography: Audio and video recording methods are employed at crime scenes less frequently than other forms of documentation. The first step in documenting a crime scene should be a crime scene video recording if one is available. If necessary, the overview and emotional analysis of the scene can be given in court. This kind of documentation should be used in conjunction with notes, sketches, and photographs, not as a replacement for them.

Pros of these Available Techniques:

- The note making is considered to be one of the handiest methods of documenting a crime scene as it includes the Name of the police station, date & time of FIR, type/nature of the crime, brief description and location of the crime scene, what evidence was collected, who collected and location of evidence found, *etc.*
- A sketch serves as a lasting record of the size and proximity of the crime scene and the tangible evidence found there. Since the other methods do not allow the spectator to gauge distances and measurements, any special information that is included in the photos or videos can also be clarified in the sketch. The simplest way to convey the measurements and layout of a crime scene is through a sketch.

- Photography is an excellent way to record a great deal of fine details, but as has already been mentioned, it should be used in conjunction with other documentation methods rather than as a replacement for them.
- Crime scene video documentation makes it possible to document the crime scene in ways that reports, interviews, and photographs cannot. A perspective on the layout of the crime scene that is not visible in images or sketches can be provided by video recording.
- The relationship between the layout of the crime scene and the evidence found there is more logical. Audio video recording allows for the documentation of a lot of details while keeping the investigator focused on their observational work. This technique of documentation is very useful in a situation where a lot of information of fluctuating or undetermined importance needs to be recorded.
- These methods of recording or documenting a crime scene are easy to use and require less efforts.

Cons of these Available Techniques:

- Chances of human error are there while making a fair sketch out of the rough one made at the crime scene.
- The chances of contamination increase with the increasing number of persons at a crime scene. Videography and photography of a crime scene require a team to walk throughout a crime scene to document the details of a crime scene.
- Documenting a crime scene *via* photography and videography needs cameras along with the ALS (Alternate Light Sources). Light source and filter cost in lakhs.
- The only other negative aspect of the audio and film documenting method is psychological. Most of the time, people just don't feel comfortable speaking *via* a tape recorder, especially if there are other people around and within hearing distance [12].

NEED OF NEW

Presently, when we talk about AR and VR, they are spreading their feet fast in forensic science. AR and VR are one of emerging fields that are being recognized in forensics. If they will be used at a crime scene for documenting it, a great change in the criminal investigation and trial system will be observed.

There are some limitations of the currently used techniques for documenting a crime scene and thus there is a need to introduce a new one. And here we are with this chapter introducing the new techniques that can be used for crime scene processing, documenting a crime scene, and preserving it for good.

At crime scenes, forensic specialists must rapidly and cleanly gather evidence. These users can study, share, and export evidence lists as well as visually tag evidence traces at crime scenes using a handheld Augmented Reality (AR) annotation tool [14]. Eight end users marked a virtual crime scene while speaking their thoughts aloud during a user walkthrough using this technology [12, 15]. Qualitative findings indicate that annotation could reduce administrative burden, hasten the data collection process, and enhance orientation to the crime scene. While the current prototype had several technical issues related to poor feature tracking, it was discovered that AR annotation was a promising, practical, and useful tool for crime scene investigation [16].

HOW WITH AR & VR?

Recent advances in the virtualization of the real environment, including potential scenarios, are of interest to the professional and scientific sectors. The use of virtual reality (VR) aims to be a reformer in terms of the innate relationship between people and technology. Virtual reality (VR) is the name given to an advanced user interface that includes a real-time simulation of the environment and allows the user to interact with it through a number of different sensory channels, including sight, hearing, touch, smell, and taste [17]. The utilization of human body sensors increases the realism of the VR experience. Health care, architecture, tactic simulators for use by the military or police, and other study areas are looking into the use of enhanced scanning and visualization techniques. This article, however, focuses on a forensic science-related application, specifically an expansion of the hotly debated subject of crime scene reconstruction [3, 4].

Virtual animations, which can depict changes over time, feature zooming or animated subjects, and/or combined photography of actual evidence (such as photos of wounds) with digitally made scenarios or characters, represent a significant advancement in this discipline. In a recent instance, a moving 3D skeleton with bullet pathways through the body was animated to show how a victim sustained their injuries. Multimodal imaging and CAD have also been used to exhibit intricate virtual representations and interactive movies of wounds and weapons [18].

Another crime investigation-related study aims to evaluate the impact of 3D visualisation in courtroom displays. The use of 3D digital data for displaying osteological evidence in courtrooms and the potential advantages of having 3D models over 2D images were extensively discussed by Errickson *et al*. The effectiveness and comprehension of 2D versus 3D display, as well as whether they are more or less prejudicial to a jury, are still unknown. According to a recent assessment, additional research is needed to fully explain and evaluate the importance of visual aids in courtroom processes.

Additionally, employing VR in criminal trials raises new concerns. Young wrote in one of his articles that VR evidence displays can be extremely compelling and even harmful, and he recommended legal authorities proceed with caution. Salmanowitz, on the other hand, asserted that VR might help lessen bias in legal judgments [7] as depicted in Fig. (1). Kilgus *et al.* issued a warning against the use of mixed reality MR and animations in courtrooms, noting the need to verify the legality, fairness, and applicability of employing visualisations for the presenting of evidence, as well as the necessity of reliable data, methodology, and visualisation. Virtual 3D models and real-world 3D reproductions have both been used as exhibits in courtrooms; nonetheless, there are high stakes and responsibilities associated with courtroom exhibitions, and certain areas call for the usage of precise definitions [19, 20]. First, it's crucial to distinguish between findings and hypothetical situations like scene reconstructions. A 3D crime scene reconstruction is based on true scientific facts as opposed to a crime scene simulation, which is based on fictitious circumstances or fictitious chains of events. The two must be distinguished when presenting 3D virtual scenes, and practitioners have a responsibility to make sure that courts of law are aware of this distinction. Second, since each has a unique set of conditions for admissibility and because demonstrative aids are not accepted as evidence and have no probative value, it is important to distinguish between exhibits used as demonstrative evidence and demonstrative aids [21].

The development of 3D capture and visualization methods recently, coupled with improvements in computer performance, has now made it possible to use VR in practical applications. Even while VR hasn't yet shown to be a trustworthy, complete tool for crime scene investigations, it's crucial to realize that 3D reconstruction and visualization methods are constantly improving. However, the court will be better able to assess the overall case if the crime scene interpretation process is more challenging. Furthermore, it appears that a smart place to start is by fusing the conventional approach with modern alternatives [3].

Although the subject of crime scene reconstruction has already been covered, it's important to make clear the difference between scene and scenario. The scene is a static, three-dimensional representation that may be viewed from multiple angles but never changes because there is no internal object movement and no process or action being demonstrated. The scenario, on the other hand, is a representation of a procedure within a synthetic environment that is built using a reconstruction of a real area. Future research may make use of tools for analyzing the physiological and psychological characteristics of people as well as the immersiveness of virtual environments [14, 22].

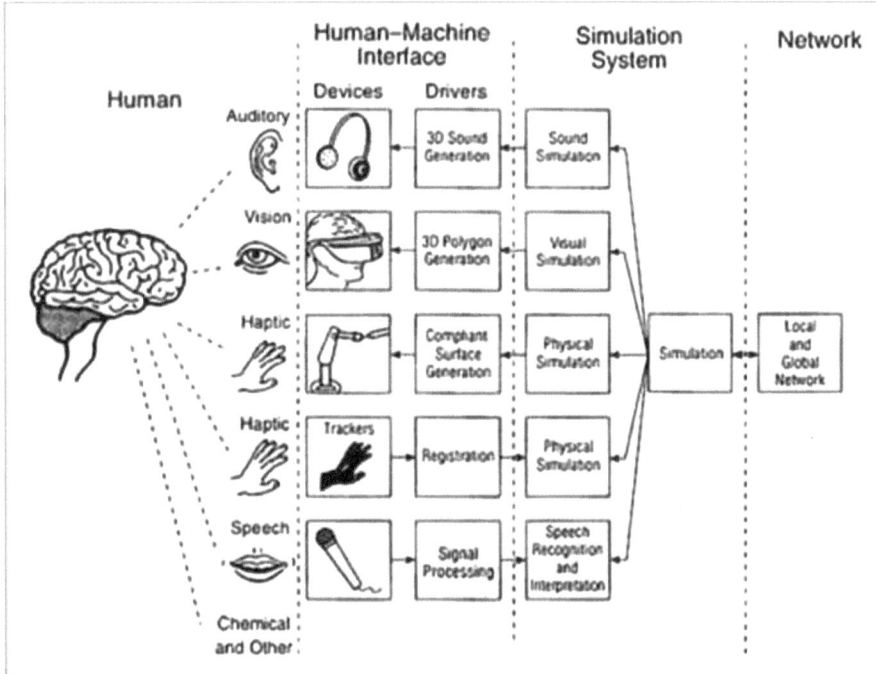

Fig. (1). Virtual Reality For Reducting Biases in Legal Process [7, 23].

Working of AR & VR: A camera-equipped device with AR software, such as a smartphone, tablet, or smart glasses, is the first step in creating an augmented reality experience. When a user looks at an object using the gadget, the software uses computer vision technology to analyse the video stream and identify it [2].

The gadget then downloads data on the object from the cloud, just like a web browser would do when loading a page from a URL. A key distinction is that instead of being given as a 2-D page on a screen, the AR information is presented as a 3-D "experience" superimposed on the item. So, the user perceives a mixture of the actual and the digital world as in Fig. (**2**) below [3].

Users may be able to interact with objects using a touchscreen, speech, or gestures while viewing real-time data flowing from them thanks to augmented reality (AR). For instance, a user may pronounce the word "stop" or tap the stop button on the digital graphic overlay within an AR experience to issue a command to a product *via* the cloud. An operator controlling an industrial robot with an AR headset might see data on the robot's performance overlay and have access to its controls [3, 4].

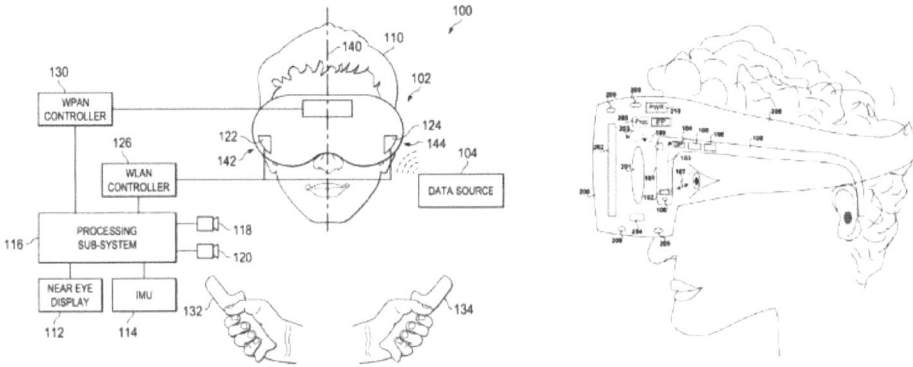

Fig. (2). AR information as a 3-D experience [3].

The size and angle of the AR display dynamically change to fit the changing context as the user moves. While older information fades out of view, new graphic or textual information appears. Users in various jobs, such as machine operators and maintenance technicians, can look at the same object in industrial settings and receive various augmented reality experiences that are catered to their needs [24].

The link between the smart item and the AR is a cloud-based 3-D digital model known as the "digital twin" of the object as shown in Fig. (**3**). Either computer-aided design, which is frequently used during product creation, or technology that digitizes real-world things are used to generate this model. In order to reflect the current reality of the product, the twin then gathers data from the product, business systems, and outside sources. It serves as the means by which the AR software precisely scales and positions current information on the object [6, 8].

Since the technology is still relatively new, it is possible that all forensic scientists won't be using VR simulations until well beyond 2018. Virtual reality has previously been used as a form of forensic evidence by some persons. A shooting scene was faithfully recreated using the Oculus Rift, according to a New Scientist story on forensic experts from the University of Zurich. This work was later confirmed by law enforcement. The opportunity to fully immerse themselves in a recreation of the crime scene would help the jury in this case to use more senses to reach a verdict. Because virtual reality is only a simulation, forensic experts would also have the freedom to alter the scene as they saw fit and remove any objects or images that might not be relevant or that viewers could find objectionable [21, 25].

Merging Real and Digital Worlds

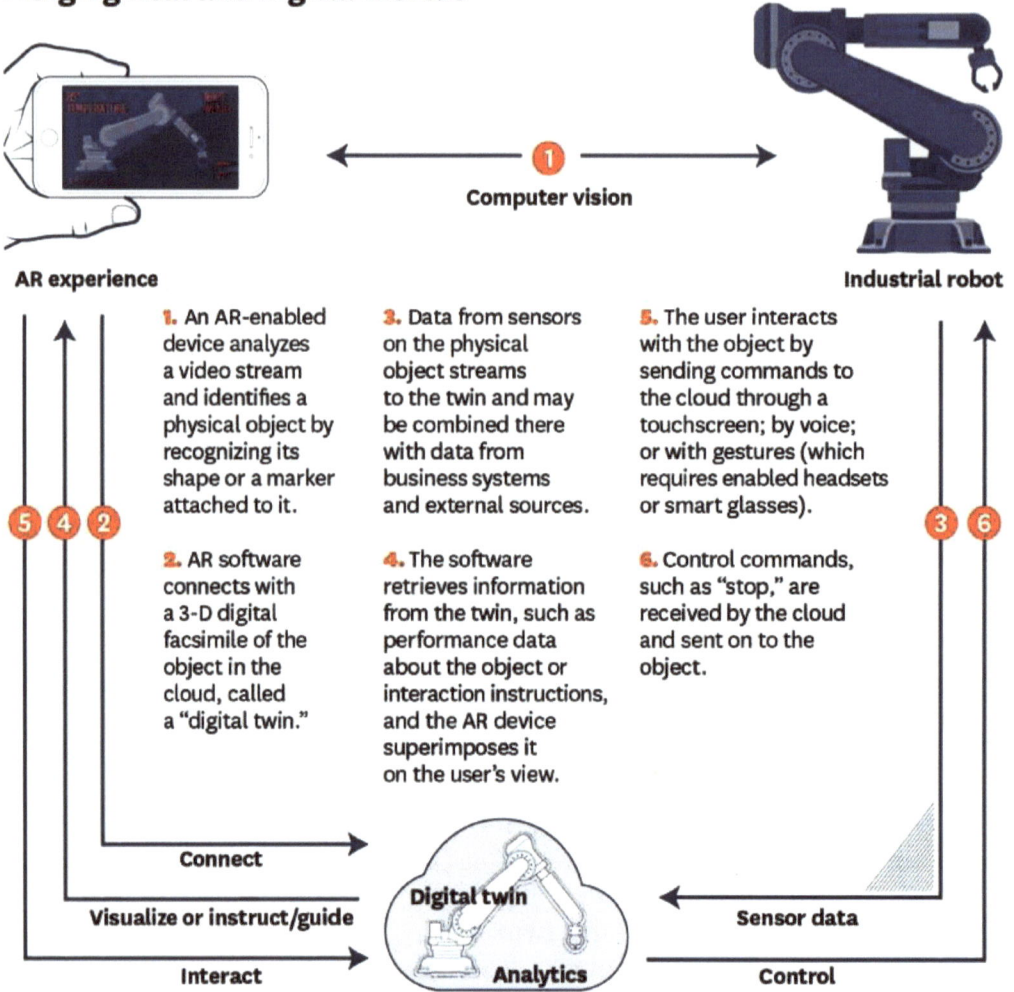

Computer vision

AR experience

Industrial robot

1. An AR-enabled device analyzes a video stream and identifies a physical object by recognizing its shape or a marker attached to it.

2. AR software connects with a 3-D digital facsimile of the object in the cloud, called a "digital twin."

3. Data from sensors on the physical object streams to the twin and may be combined there with data from business systems and external sources.

4. The software retrieves information from the twin, such as performance data about the object or interaction instructions, and the AR device superimposes it on the user's view.

5. The user interacts with the object by sending commands to the cloud through a touchscreen; by voice; or with gestures (which requires enabled headsets or smart glasses).

6. Control commands, such as "stop," are received by the cloud and sent on to the object.

Connect

Visualize or instruct/guide

Interact

Digital twin

Analytics

Sensor data

Control

ILLUSTRATION BY CLINT FORD
FROM "HOW DOES AUGMENTED REALITY WORK?"
BY MICHAEL E. PORTER AND JAMES E. HEPPELMANN, NOVEMBER–DECEMBER 2017   c HBR.ORG

Fig. (3). Digital Twin.

Virtual reality has, of course, a lot more legal use than simply that. In the future, this technology will undoubtedly be used as additional evidence in a variety of ways.

PROS AND CONS OF AR & VR

Pros:

• A Crime Scene can be documented and preserved effectively for further use.
• An interactive and virtual walkthrough of a crime scene may be given to the jury during a trial in a court of law.
• The exact location of a crime scene, condition of evidence, types/categories of evidences, dimensions, and measurements of evidence and other relevant things, *etc*. can be easily documented in the AR & VR environment.
• When done right, VR can be a thrilling sensory experience. Money and creativity are the only constraints when using computer-generated imagery (CGI) to construct alternate universes or present goods or areas in unique and entertaining ways. The Star Trek holodeck's closest equivalent is virtual reality, which will continue to advance quickly in the next years.
• When implemented correctly, augmented reality (AR) can add interesting and helpful information to a real-world situation.
• Simple augmented reality apps for smartphones are well-established.

Cons:

• AR headsets are expensive and currently available models have numerous severe flaws. Both feature a small selection of fields and using gesture controls can be challenging. Till the costs are significantly reduced and the form factor is enhanced, we won't see mainstream customer use.
• The near future will see AR headsets limited to specific applications and away from general use due to their nerdy aesthetic.

Naturally, there are drawbacks to both augmented reality and virtual reality, as with any technological advancement. The disadvantages must always be considered before a new instrument is extensively utilized, especially in the field of forensics where accuracy and care are of the utmost significance, even though in some situations, the advantages may outweigh the disadvantages [9, 17].

For instance, simulations of the murder scene from the perspective of a specific person may be created utilizing virtual reality technology. This could lead to bias among the jury in the courtroom. Damian Schofield, a digital reconstruction artist, used a murder scene as an example and was reported as saying, "Think of a murder scene: whether you view it from the point of view of the murderer, the victim, or a third person will drastically change your picture of what's happening."

Anyone employing augmented reality technology at the scene of a crime or in other circumstances may have serious concerns about distraction. Due to the additional information, they will be receiving from outside sources at other locations, law enforcement personnel employing AR technology won't be able to fully concentrate on the content in front of them. They will instead need to rely on their own skills and the knowledge of others who are guiding them through the predicament. According to one of the New Scientist articles, "since officers trialing the equipment occasionally found the new information distracting," AR technology is also inappropriate for use during making an arrest. Therefore, anyone with access to this technology must understand how and when to utilize it.

For the time being, it appears likely that tried-and-true forensic methods will continue to exist, at least in the near future, but VR and AR may provide forensic scientists and other professionals in the field with an additional tool that enables them to create a more complete picture of a crime scene.

FUTURE SCOPE

Virtual and augmented reality are without a doubt bringing significant changes on a global scale. Only ten years ago, this technology belonged to the domain of science fiction; today, it is widely used in a variety of sectors, including those related to transportation, education, and entertainment.

As technology advances swiftly, a greater number of industries are using virtual reality (VR) to achieve their goals. With more powerful tools at their disposal, parties on opposing sides of a debate can broaden their perspectives on an event, ultimately leading to more accurate judgements. Forensics is one area where VR and AR are beginning to take hold [1].

That's not to say VR technology will completely replace traditional forensic methods or be necessary in every circumstance. The applications of VR and its impacts in the courtroom are currently being tested by a number of forensic researchers.

With augmented reality (AR), the "reaction to action" cycle of first experiencing and reviewing the most recent attacks, followed by mitigating and facilitating a response, can be broken. Through the benefits of AR, police officers could be equipped to process the OODA loop more quickly to react to developing events, giving them the advantage over adversaries when handling occurrences in actual places [10, 11]. Law enforcement officers (LEAs) may deny criminals and terrorists the initiative by using AR smart glasses, while also creating a common operational picture (COP) for both the people who are on the scene and their command staff that is based on an amalgamation of data gathered from numerous sources [2, 24].

In summary, VR/AR will likely be useful in the field of forensic science, allowing these professionals to communicate with other experts more efficiently or recreate a crime scene for others to witness.

REFERENCES

[1]   A. Multani, Available at: VR/AR and the future of Forensic Technology.https://www. forensicscolleges.com/blog/resources/forensics-and-vr(2021).

[2]   J.W. Streefkerk, M. Houben, P.V. Amerongen, F.T. Haar, and J. Dijk, "The art of csi: An augmented reality tool (art) to annotate crime scenes in forensic investigation. In International conference on virtual, augmented and mixed reality", Berlin, Heidelberg. Springer, 2013.

[3]   F.S. Breien, and I. Rødseth, "Usability factors of 3D criminal archive in an augmented reality environment", *In: Proceedings of the 4th Nordic conference on Human-computer interaction: changing roles,* pp. 385-388, 2006.

[4]   A. Conway, J.I. James, and P. Gladyshev, "Development and initial user evaluation of a Virtual Crime Scene Simulator including digital evidence", In: *International Conference on Digital Forensics and Cyber Crime* Springer: Cham, 2015, pp. 16-26.
[http://dx.doi.org/10.1007/978-3-319-25512-5_2]

[5]   J. Bang, Y. Lee, Y.T. Lee, and W. Park, "AR/VR based smart policing for fast response to crimes in safe city", In: *2019 IEEE International Symposium on Mixed and Augmented Reality Adjunct (ISMAR-Adjunct)* IEEE, 2019, pp. 470-475.

[6]   K. Sandvik, and A.M. Waade, "Crime scene as augmented reality: On Screen, online and offline", *Working Papers from Crime Fiction and Crime Journalism in Scandinavia,* vol. 5, pp. 1-17, 2008.

[7]   S. Gibson, and T. Howard, "Interactive Reconstruction of Virtual Environments from Photographs", *with Application to Scene-of-Crime Analysis.,* pp. 41-48, 2000.

[8]   K. Sandvik, "Crime Scenes as Augmented Reality: Models for Enhancing Places Emotionally by Means of Narratives, Fictions and Virtual Reality", *Re-Investing Authenticity.,* Tourism, Place and Emotions, 2010.

[9]   C. Dath, *Crime scenes in Virtual Reality: A user centered study.* STOCKHOLM: SWEDEN, 2017.

[10]   D.G. Norman, K.A. Wade, M.A. Williams, and D.G. Watson, "Caught virtually lying—crime scenes in virtual reality help to expose suspects' concealed recognition", *J. Appl. Res. Mem. Cogn.,* vol. 9, no. 1, pp. 118-127, 2020.
[http://dx.doi.org/10.1016/j.jarmac.2019.12.008]

[11]   T. Sieberth, A. Dobay, R. Affolter, and L.C. Ebert, "Applying virtual reality in forensics – a virtual scene walkthrough", *Forensic Sci. Med. Pathol.,* vol. 15, no. 1, pp. 41-47, 2019.
[http://dx.doi.org/10.1007/s12024-018-0058-8] [PMID: 30519987]

[12]   "Association for Crime Scene Reconstruction", In: *Archived from the original on* ACSR, 2019.

[13]   R. Gardner, and T. Bevel, *Practical Crime Scene Analysis and Reconstruction.* CRC Press: Boca Raton, Florida, 2009, p. 1.

[14]   L. Barazzetti, R. Sala, M. Scaioni, C. Cattaneo, D. Gibelli, A. Giussani, P. Poppa, F. Roncoroni, and A. Vandone, "3D scanning and imaging for quick documentation of crime and accident scenes. Sensors, and Command, Control, Communications, and Intelligence (C3I)", *Technologies for Homeland Security and Homeland Defense XI.,* vol. 8359, pp. 1-14, 2012.

[15]   "Defining virtual reality: dimensions determining telepresence", In: *Archived from the original (PDF) on* Stanford University.: Department of Communication, 1993.

[16]   E. Agosto, A. Ajmar, P. Boccardo, F. Giulio Tonolo, and A. Lingua, "Crime scene reconstruction using a fully geomatic approach", *Sensors (Basel),* vol. 8, no. 10, pp. 6280-6302, 2008.

[http://dx.doi.org/10.3390/s8106280] [PMID: 27873870]

[17]   M. Ma, H. Zheng, and H. Lallie, "Virtual reality and 3D animation in forensic visualization", *J. Forensic Sci.,* vol. 55, no. 5, pp. 1227-1231, 2010.
[http://dx.doi.org/10.1111/j.1556-4029.2010.01453.x] [PMID: 20533989]

[18]   L.C. Ebert, T.T. Nguyen, R. Breitbeck, M. Braun, M.J. Thali, and S. Ross, "The forensic holodeck: an immersive display for forensic crime scene reconstructions", *Forensic Sci. Med. Pathol.,* vol. 10, no. 4, pp. 623-626, 2014.
[http://dx.doi.org/10.1007/s12024-014-9605-0] [PMID: 25315842]

[19]   P. S. Jasiobedzk, "Instant scene modeler for crime scene reconstruction", In: *Proceedings of IEEE Computer Society Conference on Computer Vision and Pattern Recognition* vol. 123. IEEE Computer Society: Washington, 2005, pp. 20-26.

[20]   D. Schofield, "Animating and interacting with graphical evidence: bringing courtrooms to life with virtual reconstructions", In: *Proceedings of International Conference on Computer Graphics, Imaging and Visualisation* IEEE Computer Society: Los Alamitos, CA, 2007, pp. 321-8.
[http://dx.doi.org/10.1109/CGIV.2007.18]

[21]   "How does augmented reality work? Harvard Business Review", Available at: https://hbr.org/2017/11/how-does-augmented-reality-work#:~:text=Augmented%20reality%20starts%20with%20a ,which %20analyzes%20the%20video %20stream(2021).

[22]   C. Dath, *Crime scenes in Virtual Reality: A user centered study.* STOCKHOLM: SWEDEN, 2017.

[23]   J.-L. van Gelder, M. Otte, and E.C. Luciano, "Using virtual reality in criminological research", *Crime Science,* vol. 3, no. 1, p. 10, 2014.

[24]   D. Datcu, S.G. Lukosch, and H.K. Lukosch, "Handheld augmented reality for distributed collaborative crime scene investigation", *Proceedings of the 19th International Conference on Supporting Group Work,* pp. 267-276, 2016.
[http://dx.doi.org/10.1145/2957276.2957302]

[25]   K. Sandvik, "Crime Scenes as Augmented Reality: Models for Enhancing Places Emotionally by Means of Narratives, Fictions and Virtual Reality", In: *Re-Investing Authenticity.* Tourism, Place and Emotions, 2010.

CHAPTER 6

The Influence of Green Supply Chain Management Practices Using Artificial Intelligence (AI) on Green Sustainability

S. Susithra[1], S. Vasantha[1,*] and Kabaly P. Subramanian[2]

[1] *School of Management Studies, Vels Institute of Science, Technology & Advanced Studies(VISTAS), Chennai, India*

[2] *Faculty of Business Studies, Arab Open University, Halban, Oman*

Abstract: Rapid advances in artificial intelligence (AI) are enhancing the performance of many sectors and enterprises, including green supply chain management. Innovative technologies include machine learning, IoT, and big data. AI in the manufacturing industry aims to utilise automation in production processes, better planning and forecasting, and quality products. Small and medium enterprises play a significant role in reducing carbon emissions, which has turned out to be an even more vital factor for the manufacturing industry in the past two decades. Supply chain management is one of the manufacturing's utmost areas demanding a change. Sustainable procurement enables firms to access resource recycling, efficient production, channel distribution, and end consumption to lessen their environmental impact. The 2030 Agenda for Sustainable Development (2015) is a well-thought-out synthesis of discussion that establishes sustainable growth as a critical issue for the global community. The accomplishment of sustainable goals makes it essential to develop a system of practice. This is especially important for India, which has a history of high labour intensity and industrialization. This review paper will analyse the future outlook of the market for Artificial Intelligence (AI) in GSCM and green sustainability.

Keywords: Artificial Intelligence (AI), Green Supply Chain Management, Green sustainability, SMEs.

INTRODUCTION

The concept of supply chain management has undergone multiple revolutions, evolving to reflect changes brought about by the globalisation of the economy and an elevated level of competition. It is no longer anonymous that a reliable and eff-

---
* **Corresponding author S. Vasantha:** School of Management Studies, Vels Institute of Science, Technology & Advanced Studies(VISTAS), Chennai, India; Tel: 9176132279; Fax: 91-44-22662513; E-mail: vasantha.sms@velsuniv.ac,in

Adarsh Garg, Valentina Emilia Balas, Rudra Pratap Ojha & Pramod Kumar Srivastava (Eds.)

icient supply chain management system may help a company gain a strong position in the market, and businesses grasp this notion. As a result, supply chains within businesses have grown more elastic, and corporations are continually fine-tuning their supply and demand changes for the products they deal with. Companies must compensate for many concepts from diverse domains, such as performance management, project management, talent management, quality management, and visualization of data, in order to get a competitive edge in the marketplace and maintain their financial and growth sustainability. Business dynamics and growth have been identified as the primary drivers of environmental transition [1]. Businesses are drastically altering the ecological landscape and jeopardizing the planet. Traditional corporate models do not address environmental limitations and increasingly absorbed natural resources without adhering to reusing, recycling, and re-manufacturing principles [2 - 5]. Environmental awareness has raised pressure on state and federal government organizations to evolve and effectively execute environmental safety regulations in order to prevent environmental decadence [6, 7]. It is among the primary drivers of change in the GSCM techniques [8]. Organizations have begun to integrate GSCM with other organizational activities like sourcing, manufacturing, and operations [9 - 11]. The notion of GSCM has grown in popularity as a result of knowledge sharing at numerous international conferences, and growing statistical evidence indicates a significant correlation between GSCM initiatives and company outcomes [12]. GSCM serves as a stimulus for business transitions required for a fair and sustainable economy. GSCM is defined as a system that incorporates strategic, calculative, and practical approaches for observing, assessing, and reporting the particulars of GSCM to an organization's constituents [13]. Global Supply Chain Management is a complicated process with back-an--forth movement of substance embracing the product call back, re-manufacturing, and safe disposition methods [14]. Vertical integration, which includes collaboration with customers and suppliers, promotes efficient flow in the closed curve. As a result, GSCM can be an important instrument for sustainable manufacturing and utilization of resources [15]. AI-based supply chain management solutions like computational modeling, expert systems, and agent-based systems are becoming more popular [16]. AI systems can be employed to plan, control, and manage networks in a systematic way [17].

OBJECTIVES

1. To identify the application of AI in GSCM.

2. To recognize the boons of AI usage on green sustainability.

GREEN SUPPLY CHAIN MANAGEMENT

GSCM first evolved in 1996. In reality, it is an environment protection management archetype. From the standpoint of the commodity life cycle, GSCM encompasses the stages of raw resources, goods designing, manufacturing, sales, logistics, and recycling. The organization can reduce adverse ecological impacts by achieving the best of resources and power consumption, utilizing green technology in supply chain management. The method of assessing environmental factors throughout the supply chain is known as supply chain greening. GSCM incorporates environmental needs into supply chain management at all phases of supply chain selection, greening, and product designing, which is the practice of incorporating ecological factors across the supply chain [18]. Despite the terms sustainable supply chain management (SSCM) and GSCM are often used as substitutes in the supply chain composition, the ideas subtly vary. Commercial factors, as well as social and ecological sustainability, become part and parcel of SSCM [19]. Subsequently, the notion of SSCM encompasses something beyond GSCM, and it is a subset of sustainable logistics, previously, the life of products comprised steps from designing to consumption. The environmental management method incorporates a wide range of acquisition of raw materials, design, building, usage, recovery, and the formulation of a circular system of substance flow to limit the emaciation of resources and the environmental impact [20].

Components of the Green Supply Chain

Green design implies the process of acquisition of commodities, manufacturing, and operations, knowing a complete ecosystem, people's well-being, and commodity security, with the goal of preventing environmental contamination. **Green materials** are those which make use of scarce resources and power, generate less sound, are non-toxic, and do not intoxicate the surroundings. Green productivity outperforms total management productivity. **Green production** is often referred to as clean production. Green production can be addressed in several contexts depending on the degree of development or the country. The goal of **green marketing** is to advocate the notion of sustainable growth by harmonizing the ideals of economic, ecological, and social development. **Green consumption** entails attempting to use an environment-conscious commodity and dealing with the garbage that may be hazardous to the ecological setup [21], (Fig. **1**).

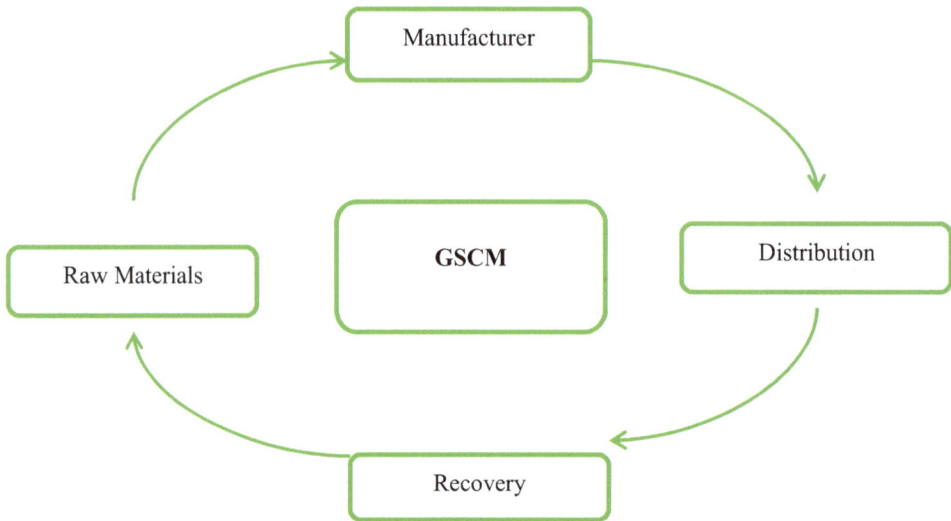

Fig. (1). Components of GSCM Source: [Nahr *et al*. 2021].

Artificial Intelligence in Green Supply Chain Management

GSCM originates from both ecological and supply chain management studies. The "green" element in the SCM entails the impact and association between the supply chain and the natural habitat. The scope of GSCM is determined by the researcher's purpose. The GSCM process is complex, and therefore, without effective blueprinting and control, the complete network might fail to achieve the expected consequences. The information from each stage of the process is required to be consolidated to mine out important information. Smart technologies can be used to collect details, and artificial intelligence could then be used to set up and operate supply chain systems to ensure long-term viability [22]. AI-based supply chain management solutions like computational modeling, and knowledge-based systems affect the GSCM systems that are included in the GSCM Technological Dimensions (GSCM TD). A green supply chain is made up of several arrangements ranging from a business process to environmental stewardship. Each arrangement is interconnected and depends on one another to complete a project.

The GSCM is an exceptionally active and sophisticated system. It encompasses a number of paradigms that require expert judgment. IoT and other technologies can be utilized to gather inputs from multiple sources, and artificial intelligence can then be used to plan and regulate GSCM systems. Through the sharing of information across vertical AI-based technology, facilitation can be a beneficial tool for connecting green consumers, vendors, and supply chain stakeholders. In green supply chain management, agent-based systems bring significant worth to the supply chain unison and collective demand forecasting. The GSCM challenges that conventional analytical models fail to handle can be solved by agent-based systems. In green logistics management, genetic algorithms can be implemented to overcome problems like vehicle scheduling, cargo loading, and material management in GSCM. Thus, artificial intelligence-based technologies influence the choice of relevant tactics and strategic plans [16, 23].

Enhancing sustainability through AI

Environmental sustainability involves decreasing the use of environmental assets and nonrenewable energy, preventing energy resource depletion, lowering waste generation and prioritizing trash reuse, reducing emissions, and recycling in multiple sectors. The use of AI to enhance the ecosystem has a significant impact on environmental preservation and defilement reduction. The application of the IoT in the subject of social sustainability promotes increased correspondence among people in regional, nationwide, and foreign bodies; it raises their consciousness and involvement with one another. This entails societal advancement. An intelligent carrier system creates a channel for different segments to connect with each other in an interactive manner by leveraging IoT technology in the process of transporting. The IoT has already given a viable and convenient foundation for intelligent transportation research. This system exchanges details concerning vehicle mobility *via* the network without the need for human interaction. As a result of the connection of things, the system's tools and equipment will become highly intelligent, such as exchanging data and actively communicating with one another. Therefore, it meets the transportation channel's goals, which will prevent accidents, rejoice passengers, and fix congestion problems [24].

Through renewable and green energy, IoT technology has a great potential to boost energy and productivity. IoT captures actual data on water and energy resources, enables better-informed resource conservation, and collects data in an easy manner to achieve congestion design and laying provisions, abate fuel use, and utilise greenhouse emissions. Therefore, sustainable development refers to the process of enhancing the environment of town life, taking into consideration the ecological, social, political, organizational, cultural, and financial divisions, without pressurizing future generations. The Internet of Things has proven its ability in the supply chain, adding up to being a breakthrough technology for all sectors. As a result, the consequences of deploying Artificial Intelligence and Internet of Things (AIoT) in the GSCM have become more apparent. Green supply chain management makes use of a wide range of AIoT applications. This improves the tracking and monitoring of products while also ensuring transparency in the organization and collaboration processes. With AIoT, the overall supply chain process may be enhanced. The key purposes of AIoT implementation in GSCM are to monitor. Factory and fleet managers can use this technology to track consignments and inventories. However, AIoT holds a greater promise for the GSCM.

The Internet of Things (IoT) provides organizations with a flow of instantaneous data regarding commodity positioning and shipping. For instance, it will be notified if the stocks are being carried in a misleading direction, and it will be feasible to track the delivery of final products or materials. Organizations can track transportation problems and identify the changes required using environmental sensors. The most widespread solution provided by AIoT is showing data regarding pressure, in-car weather conditions, humidity, *etc.,* which may harm the integrity of the commodity. These parameters can also be modified spontaneously. Companies can use AI to monitor items through transit and predict delivery, as well as anticipate and eliminate hazards associated with mishaps. The advantages of the incorporation of the GSCM range from increased storage effectiveness to improved stock management and employee welfare. For example, on-premise staff may simply discover a commodity and travel to the precise path of a given product by using GPS trackers. AIoT allows integrated productivity and effectiveness in organizations. Though, the management of AIoT is essential to make its usage effective.

Furthermore, AIoT uses artificial intelligence to fully automate the warehouse. The Internet of Things assists supply chain managers in planning for mishaps or other hindrances, by taking into account the congestion and climate. AIoT manages all information required to create prospective adaptable tools and identify the source of current lag. This system quickly gives supply chain management the required alerts. Green supply chain management is a vast topic. A significant number of processes are conducted concurrently during delivery, and managers must simplify objects using artificial intelligence of things. AIoT is one of the innovations that help the management meet the current environmental legislation and outflow limits in an intelligent distribution chain. It gives us a portrayal of how resources are being used by incorporating environmentally-sustainable practices, ecological strategies, and initiatives by using AIoT detectors for enhanced supply chain management and equity tracking.

Leveraging GSCM through Artificial Intelligence

The effects of recent technological advancements are substantial. Technological advancements in the GSCM process benefit several areas inside an organization, including human resources, sales, and marketing domains. Genetically modified algorithms, Agent-based systems, and expert systems are all examples of GSCM systems that can benefit from AI's architecture, monitoring, and management assistance as shown in Fig. (**2**). The use of AI technology might lead to the emergence of connections with eco-friendly vendors and business affiliates to improve planning efficiency. These technical components greatly aid in the sustainability of the supply chain. AI technologies such as genetic, agent-based, and modified systems are being used in the management of supply chains [25]. Artificial intelligence systems can be effectively applied in system organization, control, and management.

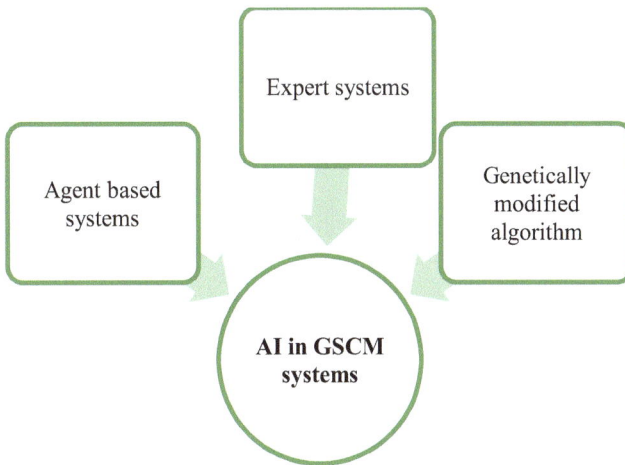

Fig. (2). AI in GSCM Systems [**Source:** Kumar *et al* 2022].

WAYS TO ACHIEVE GREEN SUSTAINABILITY THROUGH AI

Sharing Shipments has Countless Opportunities

The term "collaborative" or "sharing" transport refers to the sharing of transportation methods among several entities. This system keeps track of when shipping companies acquire and discard GPS data. As a result, the system is constantly updated with shipping circumstances, inventory loads, types of vehicles, and costs. Artificial intelligence-based technology enables businesses to share information about the supply chain with many other organizations, speed up shipments, reduce pollution, save income, and substantially green their sources.

Improved Route Planning with Autonomous Vehicles

The most significantly used logistics artificial intelligence application is the development of independent mobility. Using sophisticated GPS technologies like driverless automobiles may construct and follow avenues that are more effective than those designed by people. For instance, unmanned half-cars and rechargeable boats are currently available, and the prospect of green logistics is promising.

Planning a Delivery Quickly

Artificial intelligence maximizes shipping effectiveness by determining what is needed for shipping. AI can monitor the transportation environment to create delivery routes that balance goods and gas mileage. Likewise, AI can complete complex calculations more quickly than people.

Quick and Efficient Decision-Making

Deep reinforcement learning businesses can equip AI to make challenging and efficient choices for the green supply chain. AI can be used to specify things like the exact number of products to be shipped and the kind of vehicle to be used. AI-based systems can respond to changes and eliminate associated risks because they operate in real time. Smart devices are also used to evaluate the effectiveness of logistics management and suggest tweaks. AI can be used by businesses to locate long-term collaboration.

Savings on Fuel

Businesses implementing the Internet of Things (IoT) technology acknowledged improvements in sustainable environmental practices. However, significant changes occur naturally and cannot be rushed.

Reduced Product Waste

Businesses can actively monitor and track the conditions of things in the containers owing to artificial intelligence technology. Products can be quickly replaced if they are found to be defective, compromised, or outdated. Reverse supply chain demand can be minimized in this way because customers rarely have a need to return items.

CHALLENGES AND OPPORTUNITIES

Researchers have seen a surge in the prevalence of multinational organisations and many small-sized companies joining the hyper train in global manufacturing and marketing activities recently. With several viable sites around the world, the ambiguity of integrating eco-friendly supply chain activities expands. Embracing innovation can improve supply chain transparency, decision-making, and related activities such as internal sustainable development and enviro design for product offerings among partners from various nations. To address these issues during the implementation stage, an action plan to guide the organisation and quantification of the digitalization of GSCM is beneficial. Some organisations have established the Green supply chain innovation technology but are unsure where and how to put them into use. There are ample options to further comprehend this line of research. AI in GSCM GSC has been implemented to varying degrees in various establishments, and various factors have contributed to this. Scientific investigations can look into the factors that can affect the adoption of AI and focus more on operational processes, as many provider businesses, such as cargo handling are enabling new advancements for enhanced efficiency and ecological preservation.

CONCLUSION

GSCM has the potential to reduce the environmental effect of industrial activities while maintaining energy efficiency, cost, quality, reliability, and performance. It entails a radical shift; from edge control to complying with ecological requirements with the condition of not only minimizing environmental damage but also drives to comprehensive financial profit. The field presents numerous problems to practitioners, academics, and researchers [10]. Sophisticated technologies like AI can help organizations address GSCM-affiliated challenges in a closed economy by furnishing reliable details for convenient and adequate decision making, resulting in enhanced GSCM system knowledge management. This will improve the supply chain network's agility [26]. Firms encounter various operational issues in this changing economic climate, although these challenges mostly affect developing economies where the price is the primary concern. Implementation of an AI-based GSCM system, on the other hand, might deliver greater benefits, either explicitly or implicitly, that are far more significant in contrast to the expenditures invested during the capital phase of devising and deploying such state-of--of-art technologies.

GSCM Technological Dimensions (AI-based) have a favorable impact on GSCM approach. It has been discovered that the GSCM approach has a substantial association with GSCM processes. The GSCM approach has also been demonstrated to have a favorable impact on business performance (environmental, financial, and social). The process, however, has a greater impact on social production than it does on economic and environmental production. IoT is the most contentious problem being faced recently. Many experts believed that the Internet of Things (IoT) would be the next significant technological advancement that would triumph in the future. One such industry where the Internet of Things has proven its worth and capabilities is logistics and supply chain management. When AI and IoT are merged, the result is AIoT, or artificial intelligence of the Internet of Things. Connections expand the possibilities of smart components in products and devices by externalizing them. This enables conditions to be monitored, controlled, and optimized. AIoT not only enhances the flow of material systems but also automatic load detection and global positioning in logistics. It also improves energy efficiency, resulting in lower energy usage. In this regard, AIoT technology is a very ideal solution for automating the purposes of object recognition due to its flexibility and various benefits. By combining new AIoT technologies with other devices like sensors, it is feasible to enable continuous connection of items around us to the Internet, resulting in instant, accurate, and simple monitoring. AI enables effective optimization and network orchestration, which humans are incapable of achieving. As a result, studies on participatory decision-making systems provide a greater understanding of AI technologies, which in turn increases their proficiency. Such a technology enables AI in redefining current practices by shifting business activities from reactive to proactive, procedures, from manual to autonomous products from regular to customized, and planning from forecasting to prediction. Developments in computer microchip architecture are critical to the mainstream adoption of Artificial intelligence. As logistics are associated with transportation, using microchips for monitoring is essential. Monitoring creates an extensive amount of information that can be processed and evaluated for a variety of objectives. Mechanization of customer interactions is a new but promising field of marketing automation. Voice or virtual agents constitute a new age of client service, with great efficiency and reasonable outcomes. As these chatbots are tailored to allow for more trailblazing conversations with clients, they can be quite useful for automating client support inquiries [27].

REFERENCES

[1]   A.M. Omer, "Energy, environment and sustainable development", *Renew. Sustain. Energy Rev.,* vol. 12, no. 9, pp. 2265-2300, 2008.
        [http://dx.doi.org/10.1016/j.rser.2007.05.001]

[2]   M. Song, S. Wang, and L. Cen, "Comprehensive efficiency evaluation of coal enterprises from production and pollution treatment process", *J. Clean. Prod.,* vol. 104, pp. 374-379, 2015.
        [http://dx.doi.org/10.1016/j.jclepro.2014.02.028]

[3]   R. Dubey, A. Gunasekaran, S.J. Childe, T. Papadopoulos, S.F. Wamba, and M. Song, "Towards a theory of sustainable consumption and production: Constructs and measurement", *Resour. Conserv. Recycling,* vol. 106, pp. 78-89, 2016.
        [http://dx.doi.org/10.1016/j.resconrec.2015.11.008]

[4]   R. Dubey, "Can big data and predictive analytics improve social and environmental sustainability?", *Technol. Forecast. Soc. Change,* vol. 144, pp. 534-545, 2019.
        [http://dx.doi.org/10.1016/j.techfore.2017.06.020]

[5]   A-N. El-Kassar, and S.K. Singh, "Green innovation and organizational performance: The influence of big data and the moderating role of management commitment and HR practices", *Technol. Forecast. Soc. Change,* vol. 144, pp. 483-498, 2018.
        [http://dx.doi.org/10.1016/j.techfore.2017.12.016]

[6]   V. Mani, A. Gunasekaran, T. Papadopoulos, B. Hazen, and R. Dubey, "Supply chain social sustainability for developing nations: Evidence from India", *Resour. Conserv. Recycling,* vol. 111, pp.

42-52, 2016.
[http://dx.doi.org/10.1016/j.resconrec.2016.04.003]

[7]   M. Song, and S. Wang, "Market competition, green technology progress and comparative advantages in China", *Manage. Decis.,* vol. 56, no. 1, pp. 188-203, 2018.
[http://dx.doi.org/10.1108/MD-04-2017-0375]

[8]   Y. Li, and M. Zhang, "Green manufacturing and environmental productivity growth", *Ind. Manage. Data Syst.,* vol. 118, no. 6, pp. 1303-1319, 2018.
[http://dx.doi.org/10.1108/IMDS-03-2018-0102]

[9]   M. Song, J. Peng, J. Wang, and L. Dong, "Better resource management: An improved resource and environmental efficiency evaluation approach that considers undesirable outputs", *Resour. Conserv. Recycling,* vol. 128, pp. 197-205, 2018.
[http://dx.doi.org/10.1016/j.resconrec.2016.08.015]

[10]   S.K. Srivastava, "Green supply-chain management: A state-of-the-art literature review", *Int. J. Manag. Rev.,* vol. 9, no. 1, pp. 53-80, 2007.
[http://dx.doi.org/10.1111/j.1468-2370.2007.00202.x]

[11]   M.L. Tseng, M. Lim, K.J. Wu, L. Zhou, and D.T.D. Bui, "A novel approach for enhancing green supply chain management using converged interval-valued triangular fuzzy numbers-grey relation analysis", *Resour. Conserv. Recycling,* vol. 128, pp. 122-133, 2018.
[http://dx.doi.org/10.1016/j.resconrec.2017.01.007]

[12]   R.M. Vanalle, G.M.D. Ganga, M. Godinho Filho, and W.C. Lucato, "Green supply chain management: An investigation of pressures, practices, and performance within the Brazilian automotive supply chain", *J. Clean. Prod.,* vol. 151, pp. 250-259, 2017.
[http://dx.doi.org/10.1016/j.jclepro.2017.03.066]

[13]   R. Handfield, R. Sroufe, and S. Walton, "Integrating environmental management and supply chain strategies", *Bus. Strategy Environ.,* vol. 14, no. 1, pp. 1-19, 2005.
[http://dx.doi.org/10.1002/bse.422]

[14]   C. Hallam, and C. Contreras, "Integrating lean and green management", *Manage. Decis.,* vol. 54, no. 9, pp. 2157-2187, 2016.
[http://dx.doi.org/10.1108/MD-04-2016-0259]

[15]   S.K. Mangla, S. Luthra, N. Mishra, A. Singh, N.P. Rana, M. Dora, and Y. Dwivedi, "Barriers to effective circular supply chain management in a developing country context", *Prod. Plann. Contr.,* vol. 29, no. 6, pp. 551-569, 2018.
[http://dx.doi.org/10.1080/09537287.2018.1449265]

[16]   H. Min, "Artificial intelligence in supply chain management: Theory and applications", *Int. J. Logist.,* vol. 13, no. 1, pp. 13-39, 2010.
[http://dx.doi.org/10.1080/13675560902736537]

[17]   F. Wang, X. Lai, and N. Shi, "A multi-objective optimization for green supply chain network design", *Decis. Support Syst.,* vol. 51, no. 2, pp. 262-269, 2011.
[http://dx.doi.org/10.1016/j.dss.2010.11.020]

[18]   Nozari, Najafi, Fallah, and Lotfi, "Quantitative analysis of key performance indicators of green supply chain in fmcg industries using non-linear fuzzy method", *Mathematics,* vol. 7, no. 11, p. 1020, 2019.
[http://dx.doi.org/10.3390/math7111020]

[19]   J. Ghahremani Nahr, S.H.R. Pasandideh, and S.T.A. Niaki, "A robust optimization approach for multi-objective, multi-product, multi-period, closed-loop green supply chain network designs under uncertainty and discount", *J. Ind. Prod. Eng.,* vol. 37, no. 1, pp. 1-22, 2020.
[http://dx.doi.org/10.1080/21681015.2017.1421591]

[20]   P. Li, C. Rao, M. Goh, and Z. Yang, "Pricing strategies and profit coordination under a double echelon green supply chain", *J. Clean. Prod.,* vol. 278, p. 123694, 2021.

[http://dx.doi.org/10.1016/j.jclepro.2020.123694]

[21]   J. Ghahremani Nahr, H. Nozari, and M.E. Sadeghi, "Green supply chain based on artificial intelligence of things (AIoT)", *Int. J. Innov.Manage. Econ. Soc. Sci.,* vol. 1, no. 2, pp. 56-63, 2021.
[http://dx.doi.org/10.52547/ijimes.1.2.56]

[22]   E. Bottani, P. Centobelli, M. Gallo, M.A. Kaviani, V. Jain, and T. Murino, "Modelling wholesale distribution operations: An artificial intelligence framework", *Ind. Manage. Data Syst.,* vol. 119, no. 4, pp. 698-718, 2019.
[http://dx.doi.org/10.1108/IMDS-04-2018-0164]

[23]   S.P. Yadav, and S. Yadav, "Fusion of medical images in wavelet domain: A discrete mathematical model", In: *Ingeniería Solidaria* vol. 14. Universidad Cooperativa de Colombia- UCC., 2018, no. 25, pp. 1-11.

[24]   S.P. Yadav, D.P., Mahato, and N. T. D. Linh, *Distributed artificial intelligence.* CRC Press: Boca Raton, 2020, pp. 1-336.
[http://dx.doi.org/10.1201/9781003038467]

[25]   V. Kumar, H. Pallathadka, S. Kumar Sharma, C.M. Thakar, M. Singh, and L. Kirana Pallathadka, "Role of machine learning in green supply chain management and operations management", *Mater. Today Proc.,* vol. 51, no. 8, pp. 2485-2489, 2021.
[http://dx.doi.org/10.1016/j.matpr.2021.11.625]

[26]   M. Giannakis, and M. Louis, "A multi-agent based system with big data processing for enhanced supply chain agility", *J. Enterp. Inf. Manag.,* vol. 29, no. 5, pp. 706-727, 2016.
[http://dx.doi.org/10.1108/JEIM-06-2015-0050]

[27]   R. Toorajipour, V. Sohrabpour, A. Nazarpour, P. Oghazi, and M. Fischl, "Artificial intelligence in supply chain management: A systematic literature review", *J. Bus. Res.,* vol. 122, pp. 502-517, 2021.
[http://dx.doi.org/10.1016/j.jbusres.2020.09.009]

CHAPTER 7

Multi-Agent Based Decision Support Systems

Kuldeep Singh Kaswan[1,*], **Jagjit Singh Dhatterwal**[2] and **Ankita Tiwari**[3]

[1] *School of Computing Science & Engineering, Galgotias University, Greater Noida, India*

[2] *Department of Artificial Intelligence & Data Science, Koneru Lakshmaiah Education Foundation, Vaddeswaram, AP, India*

[3] *Department of Engineering Mathematics, Koneru Lakshmaiah Education Foundation, Vaddeswaram, AP, India*

Abstract: Multi-Agent-Based Decision Support Systems (MADSS) have emerged as powerful tools for facilitating decision-making in complex and dynamic environments. This chapter provides an overview of MADSS, highlighting their fundamental concepts, key components, and applications. MADSS leverage the principles of multi-agent systems, artificial intelligence, and decision support systems to enable collaborative decision-making among multiple autonomous agents. The chapter begins by introducing the concept of multi-agent systems, emphasizing the advantages they offer in terms of adaptability, flexibility, and scalability. It then explores the integration of decision support systems within this framework, enabling agents to make informed decisions by analyzing vast amounts of data, evaluating various alternatives, and considering multiple criteria. The architecture of MADSS is discussed, focusing on the interactions among agents, the coordination mechanisms employed, and the information exchange protocols utilized. Various agent types, such as user agents, decision agents, and knowledge agents, are described, along with their roles and responsibilities within the system. The chapter further explores the different approaches and techniques used in MADSS, including rule-based systems, expert systems, machine learning, and optimization algorithms. It highlights the importance of agent learning and adaptation to improve decision-making capabilities over time. The applications of MADSS across various domains are presented, including finance, supply chain management, healthcare, and transportation. Case studies illustrate how MADSS can enhance decision-making processes, improve efficiency, and optimize resource allocation in complex real-world scenarios.

Lastly, the chapter discusses the challenges and future directions of MADSS. Issues such as agent coordination, trust among agents, and handling uncertainty are addressed. The potential of integrating emerging technologies like blockchain, the Internet of Things (IoT), and big data analytics is also explored, envisioning more sophisticated MADSS capable of handling larger-scale problems.

---

[*] **Corresponding author Kuldeep Singh Kaswan:** School of Computing Science & Engineering, Galgotias University, Greater Noida, India; Tel: +91-9467247612; E-mail: kaswankuldeep@gmail.com

Keywords: Autonomous Agents, Decision Support System, Human Decision-making, Intelligent Decision Support Systems, Knowledge Representation, Knowledge Management, Knowledge Repositories, Multi-agent, Predictive Modeling, Software Programs.

INTRODUCTION

In today's fast-paced and interconnected world, decision-making has become increasingly complex due to the ever-growing amount of data, the dynamic nature of environments, and the involvement of multiple stakeholders. Traditional decision support systems (DSS) have provided valuable assistance in this regard by leveraging computer-based tools and techniques to aid decision-makers. However, the limitations of these systems have become apparent when faced with highly complex and uncertain situations. To address these challenges, Multi-Agent Based Decision Support Systems (MADSS) have emerged as a promising approach. MADSS combine the power of multi-agent systems (MAS), artificial intelligence (AI), and decision support systems to enable collaborative decision-making among multiple autonomous agents. By integrating the capabilities of various agents and leveraging their distributed knowledge and decision-making abilities, MADSS offer a more robust and effective solution to complex decision problems [1].

The concept of multi-agent systems lies at the core of MADSS. Multi-agent systems consist of a group of autonomous agents that interact with each other and their environment to achieve their individual goals while collectively working towards a common objective. These agents can be software entities, robots, or even humans equipped with decision-making capabilities [2]. The coordination, cooperation, and communication among these agents form the foundation of MADSS. Decision support systems, on the other hand, provide a framework for assisting decision-makers by analyzing data, evaluating alternatives, and generating insights. When combined with the principles of multi-agent systems, DSS can be extended to enable distributed decision-making, where multiple agents contribute to the decision-making process. This collaboration among agents enhances the system's ability to handle complexity, uncertainty, and the diversity of perspectives [3].

In this chapter, we delve into the realm of Multi-Agent Based Decision Support Systems. We explore the fundamental concepts underlying MADSS, including agent architectures, coordination mechanisms, and information exchange protocols. We also examine the different approaches and techniques employed within MADSS, such as rule-based systems, expert systems, machine learning, and optimization algorithms. Moreover, we discuss the wide-ranging applications of MADSS across various domains, showcasing how they have been successfully employed in finance, supply chain management, healthcare, transportation, and more [4]. Through real-world case studies, we illustrate the benefits and impact of MADSS on decision-making processes, efficiency, and resource allocation. While MADSS offer tremendous potential, they also present challenges that need to be addressed. Issues such as agent coordination, trust among agents, handling uncertainty, and scalability need careful consideration. We explore these challenges and propose potential solutions and future directions for MADSS development.

EXPERTISE OF DECISION

Expertise of decision refers to the specialized knowledge, skills, and experience possessed by individuals or systems that enable them to make effective and informed decisions in specific domains or problem areas. It encompasses the understanding of relevant concepts, principles, and patterns, as well as the ability to apply that knowledge in practical decision-making scenarios [5]. In the context of decision support systems, expertise plays a crucial role in enhancing the quality and accuracy of decisions. Decision support systems aim to capture and leverage the expertise of domain experts, either by directly incorporating their knowledge into the system or by providing tools and resources that assist decision-makers in accessing and applying that expertise. There are different types of expertise that contribute to decision-making as given below:-

Domain Expertise: This refers to the deep understanding and knowledge of a specific field or domain. Domain experts possess specialized knowledge, insights, and experience related to the subject matter, enabling them to make informed decisions within that domain. Their expertise is valuable in defining the problem space, identifying relevant factors, and evaluating potential solutions [6].

Contextual Expertise: Contextual expertise involves understanding the specific context or environment in which the decision needs to be made. It includes knowledge about the stakeholders involved, the constraints and limitations of the situation, and the potential consequences of different choices. Contextual expertise helps decision-makers consider the broader implications of their decisions and adapt their approach accordingly.

Analytical Expertise: Analytical expertise relates to the ability to gather, analyze, and interpret data and information effectively. Decision-makers with strong analytical expertise can identify patterns, extract insights, and derive meaningful conclusions from complex datasets. This expertise is particularly important in data-driven decision support systems where quantitative analysis plays a significant role [7].

Decision-Making Expertise: Decision-making expertise involves the knowledge and skills associated with the decision-making process itself. It includes understanding different decision models, techniques, and methodologies, as well as knowing how to weigh various factors and criteria when evaluating alternatives. Decision-making experts have a systematic and structured approach to decision-making, ensuring that decisions are well-informed and aligned with the desired outcomes.

In the context of decision support systems, expertise can be embedded in the system through various means. This can include the use of expert systems, which are AI-based systems that capture and mimic the decision-making processes of human experts. Expert systems employ rules, logic, and algorithms to replicate the expertise of domain experts, allowing the system to provide recommendations or insights based on that knowledge [8].

INTELLIGENT AGENTS BASED DECISION SUPPORT SYSTEM

An Intelligent Agents Based Decision Support System (IABDSS) combines the power of intelligent agents and decision support systems to provide enhanced decision-making capabilities in complex and dynamic environments. It controls the principles of artificial intelligence and multi-agent systems to create a framework where autonomous agents collaborate, communicate, and assist decision-makers in making informed choices. In an IABDSS, intelligent agents act as autonomous entities capable of perceiving their environment, reasoning, and making decisions based on predefined goals and objectives [9]. These agents possess individual expertise and knowledge, and they interact with each other and with decision-makers to collectively contribute to the decision-making process. This collaboration among agents allows for the pooling of diverse perspectives, domain expertise, and information, leading to more robust and well-informed decisions. The key components and advantages of an IABDSS include as shown below:

Intelligent Agents: These are software entities that possess decision-making capabilities, adaptability, and autonomy. Each agent has its own knowledge base, reasoning mechanisms, and ability to communicate and cooperate with other agents [10].

Decision Support Systems: The decision support system component of IABDSS provides the necessary tools, models, and algorithms to facilitate decision-making. It includes data collection, analysis, and visualization capabilities, as well as decision models and evaluation criteria.

Communication and Coordination Mechanisms: Effective communication and coordination mechanisms enable agents to exchange information, negotiate, and collaborate towards a common goal. These mechanisms can include protocols, shared databases, message passing, or other forms of inter-agent communication [11].

Knowledge Management: IABDSS incorporates mechanisms for knowledge sharing, acquisition, and representation. This allows agents to access and utilize relevant information and expertise, including domain-specific knowledge, historical data, and expert opinions.

Enhanced Decision-Making: By leveraging the expertise and intelligence of multiple agents, IABDSS can provide more comprehensive and accurate decision support. Agents can analyze vast amounts of data, evaluate alternatives, and consider various criteria to generate insights and recommendations [12].

Adaptability and Scalability: Intelligent agents are capable of adapting to changing environments and can scale the decision support system to handle complex and dynamic problems. As the system grows, more agents can be added to provide additional expertise and handle increasing amounts of data [13].

Collaboration and Consensus Building: IABDSS facilitates collaboration among agents, promoting consensus building and collective decision-making. Agents can share information, negotiate, and reconcile different viewpoints, leading to more robust and well-rounded decisions.

Efficient Resource Allocation: Through the intelligent allocation of resources, IABDSS can optimize resource utilization and allocation. Agents can consider constraints, preferences, and objectives to suggest the most effective allocation strategies [14].

CONCEPT OF DECISION TREES

Decision trees are a popular and intuitive machine-learning technique used for classification and regression tasks. They provide a visual representation of decision-making processes by organizing data into a hierarchical structure resembling a tree. Each node in the tree represents a decision based on a particular feature or attribute, leading to subsequent nodes or leaf nodes that represent the final decision or outcome [15]. The fundamental concept behind decision trees is to partition the data based on features and make decisions by evaluating the attributes that best separate the data into distinct classes or groups. The decision tree algorithm learns from labeled training data to construct an optimal tree structure that can generalize well to new, unseen data. The key components and advantages of a decision tree include shown as given below:

Root Node: The topmost node of the tree represents the initial decision based on a selected attribute. It divides the dataset into subgroups based on different attribute values.

Internal Nodes: These nodes represent intermediate decisions based on specific attributes. Each internal node corresponds to a question or condition that guides the splitting of the data into further subgroups [16].

Branches: The branches represent the possible outcomes or values of the attribute at each node. They connect the nodes, leading to subsequent nodes or leaf nodes.

Leaf Nodes: Also known as terminal nodes, leaf nodes represent the final decision or outcome of the decision-making process. Each leaf node corresponds to a specific class label or regression value [17].

The construction of a decision tree involves selecting the most informative attributes or features at each node to maximize the separation or predictive power. This is typically achieved by employing measures such as entropy, information gain, or Gini index, which quantify the impurity or disorder of the data and decision trees offer several advantages:

Interpretability: Decision trees provide a transparent and interpretable representation of decision-making. The tree structure allows decision paths to be easily understood, enabling users to gain insights into the decision process [18].

Handling Both Categorical and Numerical Data: Decision trees can handle both categorical and numerical attributes, making them versatile for various types of data.

Robustness to Outliers and Irrelevant Features: Decision trees are relatively robust to outliers and can handle irrelevant features without significantly impacting their performance.

Scalability: Decision trees can efficiently handle large datasets and can be parallelized, making them suitable for scalable machine learning applications.

KNOWLEDGE-BASED DECISION TREE

A knowledge-based decision tree is an extension of the traditional decision tree algorithm that incorporates expert knowledge or domain-specific rules into the decision-making process. It combines the power of decision trees with the expertise of human knowledge, enabling more accurate and informed decisions [19]. In a knowledge-based decision tree, the construction of the tree involves not only the analysis of the training data but also the incorporation of predefined rules or knowledge provided by domain experts. These rules are typically represented as additional conditions or constraints at specific nodes of the decision tree. The knowledge-based approach aims to enhance the decision tree's performance by guiding the decision-making process based on expert insights and domain-specific considerations [20]. The integration of expert knowledge brings several benefits to decision tree algorithms given below:

Improved Accuracy: By incorporating domain-specific rules, a knowledge-based decision tree can make more accurate decisions. Expert knowledge helps to capture important factors, exceptions, or special cases that may not be evident from the data alone. This improves the overall performance and reliability of the decision tree [21].

Interpretability: Like traditional decision trees, knowledge-based decision trees offer interpretability. The explicit incorporation of expert rules allows decision paths to be easily understood and validated. This transparency enables users to gain insights into the decision-making process and have confidence in the decisions made by the system.

Handling Complex or Incomplete Data: Expert rules can help fill gaps or deal with missing values in the data. By leveraging expert knowledge, the decision tree can make reliable decisions even when faced with incomplete or imperfect data. The rules act as a guide to compensate for data limitations and provide accurate predictions or classifications.

Incorporation of Business Constraints: Knowledge-based decision trees allow the inclusion of business rules or constraints that reflect real-world limitations or requirements. This ensures that the decision-making process aligns with organizational policies, legal regulations, or specific business needs [22].

The process of constructing a knowledge-based decision tree involves integrating the expert rules into the decision tree algorithm. This can be done by modifying the splitting criteria at specific nodes or by introducing additional conditions that must be satisfied for a particular decision path. The rules can be defined in various forms, including logical expressions, decision tables, or rule-based systems.

XML BASED KNOWLEDGE REPRESENTATION TECHNOLOGY

XML (Extensible Markup Language) is a widely used technology for representing and structuring data in a hierarchical format. It provides a flexible and standardized approach to encoding information, making it suitable for knowledge representation in various domains. XML-based knowledge representation technology leverages XML to capture and store knowledge in a structured and machine-readable format [23]. XML-based knowledge representation technology encompasses various techniques and approaches given below:

XML Schema: XML Schema is a specification language that allows for defining the structure, data types, and constraints of XML documents. It can be used to create schemas that define the specific knowledge representation model, ensuring data consistency and validity.

Ontology Representation: XML can be used to represent ontologies, which capture concepts, relationships, and rules in a specific domain. Ontologies expressed in the XML format allow for the representation of classes, properties, instances, and axioms, enabling sophisticated knowledge modeling and reasoning [24].

Semantic Annotations: XML can be used to annotate textual content with semantic information. XML-based annotations provide a way to represent knowledge embedded in text, such as concepts, relationships, and metadata. This facilitates information retrieval, extraction, and semantic search.

Knowledge Markup Languages: XML-based markup languages, such as the Knowledge Markup Language (KML) or the Darwin Information Typing Architecture (DITA), provide specialized frameworks for representing knowledge in specific domains. These languages define a set of tags, rules, and structures tailored to capture and represent knowledge effectively.

SOFTWARE AGENTS-BASED KNOWLEDGE REPRESENTATION

Software agents-based knowledge representation is an approach that allows intelligent software agents to represent and manipulate knowledge in a distributed and autonomous manner. This technology combines the principles of artificial intelligence, knowledge representation, and multi-agent systems to enable efficient and collaborative knowledge management. In this approach, software agents act as autonomous entities with the ability to perceive their environment, reason, and communicate with other agents. Each agent possesses its own knowledge base, which stores and represents domain-specific knowledge. The knowledge can be encoded using various formalisms such as ontologies, rules, or semantic networks [25].

Software agents facilitate the representation of knowledge by actively gathering and sharing information, reasoning over it, and making decisions based on the knowledge they possess. They can collaborate with other agents, exchanging knowledge and combining their expertise to achieve shared goals or solve complex problems. This collaborative knowledge representation enables effective knowledge sharing, integration, and utilization across distributed systems. The use of software agents in knowledge representation offers several advantages. Firstly, it enables the distribution of knowledge across multiple agents, making it possible to handle large-scale and decentralized knowledge management tasks. Agents can autonomously acquire and update knowledge from diverse sources, ensuring a dynamic and up-to-date representation. Secondly, software agents provide a platform for intelligent knowledge reasoning and decision-making. Agents can use their knowledge representation capabilities to perform complex reasoning tasks, such as inferencing, pattern recognition, or decision support. They can integrate different types of knowledge and apply various reasoning mechanisms to derive meaningful insights or make informed decisions [26].

Furthermore, software agents-based knowledge representation promotes adaptability and flexibility. Agents can autonomously adapt to changing environments, update their knowledge bases, and adjust their behavior based on new information or requirements. This dynamic nature allows knowledge representation to be more responsive and agile in dynamic and evolving domains.

INTELLIGENT BASED DECISION SUPPORT ARCHITECTURE

An intelligent-based decision support architecture refers to a framework that combines intelligent technologies, such as artificial intelligence (AI) and machine learning, with decision support systems (DSS) to enhance the decision-making process. This architecture leverages advanced computational techniques and algorithms to provide intelligent insights, recommendations, and predictions, enabling decision-makers to make more informed and effective choices [27]. At the core of an intelligent-based decision support architecture is the integration of intelligent components, such as data analytics, knowledge representation, and reasoning mechanisms. These components work together to analyze large volumes of data, extract patterns, and generate meaningful insights. They also incorporate domain-specific knowledge, expert rules, and statistical models to facilitate accurate decision-making. The architecture typically includes the following key elements:

Data Integration: This component focuses on collecting and integrating data from various sources, both internal and external to the organization. It involves data preprocessing, cleansing, and transformation to ensure data quality and consistency.

Knowledge Representation: The architecture incorporates mechanisms to represent and organize domain-specific knowledge, including rules, ontologies, and semantic networks. This allows for the structured representation of knowledge and facilitates reasoning and inference processes.

Analytics and Modeling: This component employs advanced analytics techniques, such as data mining, machine learning, and predictive modeling, to uncover patterns, trends, and relationships in the data. These techniques enable the identification of insights and the generation of predictions or recommendations [28].

Decision Support Engines: The decision support engines leverage the analytical outputs and knowledge representation to provide intelligent decision support. They use algorithms and models to evaluate alternatives, assess risks, and generate recommendations based on predefined criteria or user preferences.

Visualization and User Interface: The architecture includes user-friendly interfaces and visualizations to present the analyzed data, insights, and recommendations to decision-makers. This enhances understanding and facilitates effective decision-making by providing clear and intuitive representations of complex information.

LEARNING AGENTS

Learning agents are intelligent software entities that have the ability to acquire knowledge and improve their performance through experience. These agents use learning algorithms and techniques to analyze data, identify patterns, and adapt their behavior based on the observed feedback or outcomes in Fig. (1) The goal of learning agents is to autonomously learn from their interactions with the environment and make better decisions over time [29].

Knowledge Agents

Knowledge agents are intelligent software entities that are designed to manage and manipulate knowledge within a system or organization. These agents are specifically focused on acquiring, organizing, representing, and utilizing knowledge to support decision-making, problem-solving, and other cognitive tasks. Knowledge agents play a crucial role in knowledge management systems by leveraging their capabilities to gather information, extract insights, and represent knowledge in a structured and meaningful way. They can perform tasks such as knowledge acquisition, knowledge representation, knowledge integration, and knowledge dissemination. Knowledge acquisition involves the process of gathering relevant information from various sources, such as databases, documents, or external systems. Knowledge agents can employ techniques like web scraping, natural language processing, or data mining to collect and extract knowledge from unstructured or semi-structured data sources [30].

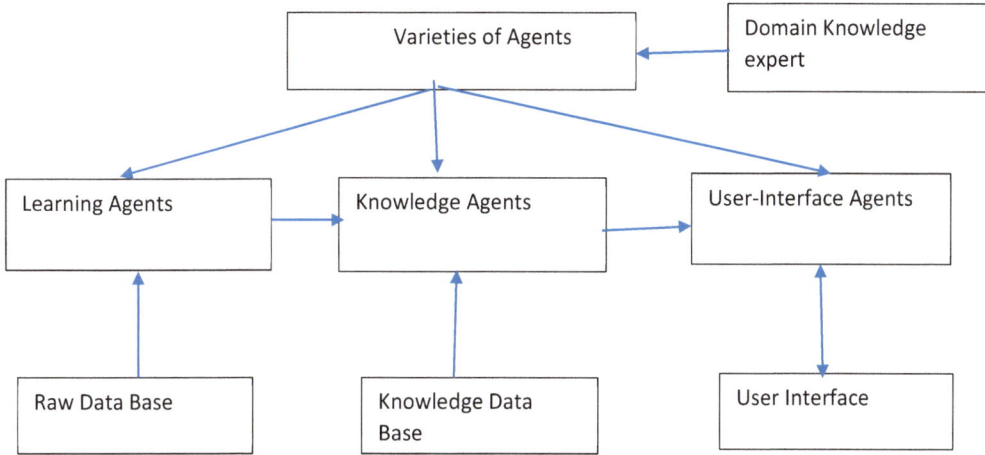

Fig. (1). Agents based Domain Knowledge specific.

User Agents

User agents, also known as personal agents or intelligent user interfaces, are software entities that act on behalf of users to perform specific tasks or assist in interactions with computer systems. These agents are designed to understand user preferences, anticipate user needs, and autonomously perform actions or provide personalized information and recommendations. User agents aim to enhance user experience and simplify complex interactions with computer systems by leveraging artificial intelligence and machine learning techniques. They can adapt to user behavior, learn from user interactions, and customize their functionalities to better serve individual users [31].

IMPLEMENTATION OF THE ID3

Implementation of the ID3 algorithm involves the practical application and execution of the ID3 (Iterative Dichotomiser 3) decision tree algorithm, a popular and widely used machine learning technique for classification tasks. The ID3 algorithm constructs a decision tree by iteratively selecting the most informative attribute as the root node and recursively splitting the data based on the values of that attribute. The implementation of ID3 involves several steps, including data preprocessing, attribute selection, tree construction, and tree pruning. During data preprocessing, the input data is prepared by handling missing values, handling categorical attributes, and normalizing or scaling numeric attributes. The attribute selection step utilizes measures such as entropy or information gain to determine the most informative attribute for splitting the data. The selected attribute becomes the root node of the decision tree, and the data is partitioned into subsets based on its distinct attribute values. This process is recursively applied to each subset until a stopping condition is met, such as reaching a pure class or a predefined depth limit. Additionally, to avoid overfitting, tree pruning techniques may be applied to remove unnecessary branches or nodes. The implementation of the ID3 algorithm requires careful consideration of various aspects, such as data preparation, attribute selection, tree construction, and pruning, to ensure accurate and effective classification results [32].

Business Application

The implementation of the ID3 algorithm in business applications can bring several benefits and support decision-making processes. Here are a few examples of business applications where ID3 can be applied:

Customer Segmentation: ID3 can be used to segment customers based on various attributes such as demographics, purchasing behavior, or preferences. By constructing a decision tree using the ID3 algorithm, businesses can identify distinct customer segments and tailor marketing strategies, product recommendations, and personalized offers to each segment.

Fraud Detection: ID3 can be utilized to detect fraudulent activities or transactions by analyzing patterns and anomalies in data. The algorithm can learn from historical data that contains labeled instances of fraud and non-fraud cases, enabling the creation of a decision tree that can effectively identify suspicious patterns and flag potential fraudulent activities.

Risk Assessment: ID3 can assist businesses in assessing risks associated with certain decisions or scenarios. By constructing a decision tree based on historical data and relevant attributes, businesses can analyze the likelihood and impact of different risk factors. This enables informed decision-making and proactive risk management strategies.

Employee Performance Evaluation: ID3 can be used to evaluate employee performance by analyzing various factors such as skills, experience, and performance metrics. By constructing a decision tree, businesses can identify key attributes that contribute to high-performing employees, enabling better talent management and targeted training programs.

Supply Chain Optimization: ID3 can help optimize supply chain operations by identifying critical factors and decision points. By constructing a decision tree based on historical data and attributes such as demand, inventory levels, and delivery times, businesses can make informed decisions regarding procurement, production scheduling, and logistics planning.

Credit Scoring: ID3 can assist in credit scoring and risk assessment for lending institutions. By analyzing customer data, financial indicators, and credit history, the algorithm can construct a decision tree to assess creditworthiness and predict default probabilities. This enables lenders to make accurate and fair decisions when evaluating loan applications.

CONCLUSION

Multi-Agent-Based Decision Support Systems (MADBSS) have emerged as a promising approach to enhancing decision-making processes in complex and dynamic environments. By leveraging the power of intelligent software agents, MADBSS provide a collaborative and distributed framework that combines the expertise, knowledge, and decision-making capabilities of multiple agents. This chapter has explored the key aspects and benefits of MADBSS. MADBSS offer several advantages over traditional decision support systems. They enable the integration of diverse perspectives and domain expertise by allowing agents to share and exchange knowledge, collaborate, and contribute to the decision-making process. The autonomous nature of agents empowers them to adapt to changing environments, acquire and analyze vast amounts of data, and make informed decisions based on their individual and collective intelligence. The use of intelligent agents in decision support systems also improves the accuracy, efficiency, and effectiveness of decision-making. Agents can employ advanced computational techniques, machine learning algorithms, and expert knowledge to analyze complex data, identify patterns, and generate intelligent insights and recommendations. The collaborative nature of MADBSS facilitates consensus-building, mitigates biases, and enhances the quality and robustness of decisions.

Moreover, MADBSS enable decision-making in real-time or near-real-time scenarios, where timely and accurate decisions are crucial. The distributed nature of the architecture allows for parallel processing, scalability, and handling of large-scale and complex decision problems. The integration of intelligent agents with decision support systems provides a flexible and adaptable platform for addressing diverse decision challenges across various domains. Despite the significant benefits, the adoption and implementation of MADBSS come with challenges. These include issues related to data privacy and security, ensuring interoperability and communication among agents, and designing effective coordination and negotiation mechanisms. Additionally, the development and maintenance of intelligent agents require expertise in AI, machine learning, and knowledge engineering.

REFERENCES

[1]   P. Adriaans, and D. Zantinge, *Data mining* Addison-Wesley: Harlow, UK, 1996.

[2]   P. Chithaluru, A. Singh, J.S. Dhatterwal, A.H. Sodhro, M.A. Albahar, A. Jurcut, and A. Alkhayyat, "An optimized privacy information exchange schema for explainable AI empowered wimax-based IoT networks", *Future Gener. Comput. Syst.,* vol. 148, pp. 225-239, 2023.
[http://dx.doi.org/10.1016/j.future.2023.06.003]

[3]   S. Ba, K.R. Lang, and A.B. Whinston, "Enterprise decision support using Intranet technology", *Decis. Support Syst.,* vol. 20, no. 2, pp. 99-134, 1997.
[http://dx.doi.org/10.1016/S0167-9236(96)00068-1]

[4]   JS Dhatterwal, S Dixit, and S Srinivasan, "Implementation of case base reasoning system using multi-agent system technology for a buyer and seller negotiation system", *Int. J. of Modern Elec. and Commu. Eng.,* vol. 7, no. 3, pp. 63-67, 2019.

[5]   N. Bolloju, M. Khalifa, and E. Turban, "Integrating knowledge management into enterprise environments for the next generation decision support", *Decis. Support Syst.,* vol. 33, no. 2, pp. 163-176, 2002.
[http://dx.doi.org/10.1016/S0167-9236(01)00142-7]

[6]   K. S. Kaswan, J. S. Dhatterwal, S. Grima, and K. Sood, Robotic process automation applications area in the financial sector. *In: Intelligent multimedia technologies for financial risk management: Trends, tools and applications.* Process Automation, 2023, pp. 279-296.
[http://dx.doi.org/10.1049/PBPC060E_ch13]

[7]   S Srinivasan, "Multi-agent-based decision support system using data mining and case based reasoning", *Int. J. of Computer Sci. Issues.,* vol. 8, no. 4, 2011.

[8]   S.P. Yadav, and S. Yadav, Fusion of medical images in wavelet domain: A discrete mathematical model.*Solidar. Eng.,* vol. 14, no. 25, pp. 1-11, 2018.
[http://dx.doi.org/10.16925/.v14i0.2236]

[9]   U. Fayyad, G. Piatetsky-Shapiro, and P. Smyth, From data mining to knowledge discovery: an overview.*Advances in knowledge discovery and data mining.,* U. Fayyad, G. Piatetsky-Shapiro, P. Smyth, R. Uthurusamy, Eds., AAAI/MIT Press: Cambridge, MA, 1996, pp. 1-36.

[10]   K. Kuldeep Singh, "Intelligent agents based integration of machine learning and case base reasoning system", In: *2nd International Conference on Advance Computing and Innovative Technologies in Engineering (ICACITE)*Greater Noida, India, 2022, pp. 28-29.
[http://dx.doi.org/10.1109/ICACITE53722.2022.9823890]

[11]   V. Vashisht, A.K. Pandey, and S.P. Yadav, Speech recognition using machine learning. *IEIE Trans. Smart Process. Comput.,* vol. 10, no. 3, pp. 233-239, 2021.
[http://dx.doi.org/10.5573/IEIESPC.2021.10.3.233]

[12]   C.W. Holsapple, and M. Singh, "Toward a unified view of electronic commerce, electronic business, and collaborative commerce: A knowledge management approach", *Knowl. Process Manage.,* vol. 7, no. 3, pp. 151-164, 2000.
[http://dx.doi.org/10.1002/1099-1441(200007/09)7:3<151::AID-KPM83>3.0.CO;2-U]

[13]   S.C. Hui, and G. Jha, "Data mining for customer service support", *Inf. Manage.,* vol. 38, no. 1, pp. 1-13, 2000.
[http://dx.doi.org/10.1016/S0378-7206(00)00051-3]

[14]   JS Dhatterwal, S Dixit, and S Srinivasan, "The role of mas based cbrs using dm techniques for the supplier selection", *Int. j. comput. sci. eng.,* vol. 7, no. 5, pp. 1658-1665, 2019.

[http://dx.doi.org/10.26438/ijcse/v7i5.16581665]

[15] M.Y. Kiang, "A comparative assessment of classification methods", *Decis. Support Syst.,* vol. 35, no. 4, pp. 441-454, 2003.
[http://dx.doi.org/10.1016/S0167-9236(02)00110-0]

[16] C.N. Kim, H. Michael Chung, and D.B. Paradice, "Inductive modeling of expert decision making in loan evaluation: A decision strategy perspective", *Decis. Support Syst.,* vol. 21, no. 2, pp. 83-98, 1997.
[http://dx.doi.org/10.1016/S0167-9236(97)00022-5]

[17] Y. Kudoh, M. Haraguchi, and Y. Okubo, "Data abstractions for decision tree induction", *Theor. Comput. Sci.,* vol. 292, no. 2, pp. 387-416, 2003.
[http://dx.doi.org/10.1016/S0304-3975(02)00178-0]

[18] J. Mao, and I. Benbasat, "The use of explanations in knowledge-based systems: Cognitive perspectives and a process-tracing analysis", *J. Manage. Inf. Syst.,* vol. 17, no. 2, pp. 153-179, 2000.
[http://dx.doi.org/10.1080/07421222.2000.11045646]

[19] S. McIlraith, T. C. Son, and H. Zeng, "Semantic web services", *Intell. Syst.,* vol. 16, no. 53, pp. 153-179, 2001.

[20] Jagjit Singh Dhatterwal, and Kuldeep Singh Kaswan, Intelligent agent based case base reasoning systems build knowledge representation in COVID-19 analysis of recovery infectious patients. *Applications of Artificial Intelligence in COVID-19 . Medical Virology: From Pathogenesis to Disease Control.,* S. Mohanty Nandan, S.K. Saxena, S Satpathy, J.M. Chatterjee, Eds., Springer: Singapore, 2020.
[http://dx.doi.org/10.1007/978-981-15-7317-0]

[21] B. Padmanabhan, and A. Tuzhilin, "Unexpectedness as a measure of interestingness in knowledge discovery", *Decis. Support Syst.,* vol. 27, no. 3, pp. 303-318, 1999.
[http://dx.doi.org/10.1016/S0167-9236(99)00053-6]

[22] J.R. Quinlan, "Improved use of continuous attributes in C4.5", *J. Artif. Intell. Res.,* vol. 4, no. 4, pp. 77-90, 1996.
[http://dx.doi.org/10.1613/jair.279]

[23] J.R. Quinlan, "Learning first-order definitions of functions", *J. Artif. Intell. Res.,* vol. 5, no. 5, pp. 139-161, 1996.
[http://dx.doi.org/10.1613/jair.308]

[24] J.P. Shim, M. Warkentin, J.F. Courtney, D.J. Power, R. Sharda, and C. Carlsson, "Past, present, and future of decision support technology", *Decis. Support Syst.,* vol. 33, no. 2, pp. 111-126, 2002.
[http://dx.doi.org/10.1016/S0167-9236(01)00139-7]

[25] Jagjit Singh Dhatterwal, and Kuldeep Singh Kaswan, Intelligent agent based case base reasoning systems build knowledge representation in COVID-19 analysis of recovery infectious patients. *Applications of Artificial Intelligence in COVID-19 . Medical Virology: From Pathogenesis to Disease Control.,* S. Mohanty Nandan, S.K. Saxena, S Satpathy, J.M. Chatterjee, Eds., Springer: Singapore, 2020.
[http://dx.doi.org/10.1007/978-981-15-7317-0]

[26] A.C. Stylianou, G.R. Madey, and R.D. Smith, "Selection criteria for expert systems shells: A socio-technical framework", *Commun. ACM.,* vol. 10, no. 35, pp. 30-48, 1992.
[http://dx.doi.org/10.1145/135239.135240]

[27] T.K. Sung, N. Chang, and G. Lee, "Dynamics of modeling in data mining: Interpretive approach to bankruptcy prediction", *J. Manage. Inf. Syst.,* vol. 16, no. 1, pp. 63-85, 1999.
[http://dx.doi.org/10.1080/07421222.1999.11518234]

[28] Y. Shoham, "Agent-oriented programming", *Artif. Intell.,* vol. 60, no. 1, pp. 51-92, 1993.
[http://dx.doi.org/10.1016/0004-3702(93)90034-9]

[29] E. Takimoto, and A. Maruoka, "Top-down decision tree learning as information based boosting",

Theor. Comput. Sci., vol. 292, no. 2, pp. 447-464, 2003.
[http://dx.doi.org/10.1016/S0304-3975(02)00181-0]

[30]   M.S. Tsechansky, N. Pliskin, G. Rabinowitz, and A. Porath, "Mining relational patterns from multiple relational tables", *Decis. Support Syst.,* vol. 27, no. 1-2, pp. 177-195, 1999.
[http://dx.doi.org/10.1016/S0167-9236(99)00043-3]

[31]   A. Whinston, "Intelligent agents as a basis for decision support systems", *Decis. Support Syst.,* vol. 20, no. 1, p. 1, 1997.
[http://dx.doi.org/10.1016/S0167-9236(96)00071-1]

[32]   R.O. Weber, and D.W. Aha, "Intelligent delivery of military lessons learned", *Decis. Support Syst.,* vol. 34, no. 3, pp. 287-304, 2003.
[http://dx.doi.org/10.1016/S0167-9236(02)00122-7]

CHAPTER 8

An Artificial Intelligence Integrated Irrigation System: A Smart Approach

Vibhooti Narayan Mishra[1,*], Divya Pratap Singh[2] and Radheshyam Dwivedi[3]

[1] *Department of Mechanical Engineering, NIT Patna, Bihar, India*

[2] *Department of Applied Sciences and Humanities, Rajkiya Engineering College, Azamgarh, Uttar Pradesh, India*

[3] *Department of Electrical Engineering, MNNIT Allahabad, UP, India*

Abstract: The farming sector is considered the backbone of the Indian economy. The demand for water is continuously increasing with rising population density. Water is frequently wasted on the land due to unscientific irrigation techniques and unpredictable weather conditions. The efficiency of irrigation networks is challenged by the extremely variable and farmer-dependent irrigation water demand. Each farm's irrigation intensity is influenced by both accurate and inaccurate variables, as well as the farmer's behaviour. Accurate and inaccurate variables include soil moisture, crop's water requirement, and climate conditions. An auto solar-powered smart irrigation system enables users to accurately time watering cycles by tracking the soil moisture at numerous sites across the field. This system also brings down the utilization of grid power to save electricity as per the energy crisis for Indian farmers. The objective of our work is to use an automated watering system to reduce the farmer's manual involvement in the field at an effective cost. The artificial intelligence (AI) system is based on sensing a control mechanism with required correction for the maximum yielding of irrigation. It also optimizes the water requirement of a variety of crops. A more accessible and more affordable solution to this issue is provided by the present work. The conventional methods of irrigation used in India are sprinklers and flood-type systems. A large amount of water gets wasted, and crops are destroyed due to the uneven slopes of the field. These problems can be resolved by incorporating an intelligent automated irrigation system with an AI.

Keywords: Artificial Intelligence, Solar Energy, Irrigation, Microcontroller, Smart System.

INTRODUCTION

World's population is increasing day by day and is predicted to be 10 billion by the end of 2050. Agriculture is playing an important role in employment

---

[*] **Corresponding author Vibhooti Narayan Mishra:** Department of Mechanical Engineering, NIT Patna, Bihar, India; E-mail: vibhooti1810@gmail.com

Adarsh Garg, Valentina Emilia Balas, Rudra Pratap Ojha & Pramod Kumar Srivastava (Eds.)

generation to boost the Indian economy. The previous two years have seen a brisk expansion in the agriculture industry. The industry, which employs the most people, contributed a significant 18.8% (2021–2022) of the country's Gross Value Added (GVA), growing by 3.6% in 2020–2021 and 3.9% in 2021–2022. Approximately 38% of the land of the world is used for crop production purposes. Par capita income of the rural economy is mostly dependent on agriculture. A network of big and smaller canals from rivers, underground water systems, dams, and rainwater installations are all part of India's irrigation infrastructure [1]. Groundwater water systems are conventionally used in India, and the water level is exhausted day by day. 39-million-hectare land out of 160 million hectares is irrigated using a groundwater system. Two hundred twenty-five of the land is irrigated using a large and small canal irrigation system. The government aims to increase the canal network to enhance irrigation using canals [2]. The irrigation of 67% of cultivated land in India totally depends on the uncontrolled monsoon [3]. Due to the uneven availability of water, only 35% of the total land in India is irrigated only. As whole countries in the world are facing climate change, monsoon in India is also not periodic over the year. This problem leads to the destruction of crops either due to a shortage of water or plenty of water [4]. Therefore, Indian farmers need a smart and intelligent mechanism to control the humidity of soil so that the yield can be increased. Initially, Artificial intelligence was incorporated with fuzzy logic to explain human behaviour. Artificial intelligence predicts water demand using the ground truth and historical data incorporation of fuzzy logic and other intelligent techniques [5].

The use of AI-based technology helps to increase productivity across all industries, including the agricultural sector, and to address difficulties like crop disease, water wastage, agriculture, weeding, and harvesting. Agriculture yield can be enhanced using AI techniques apart from fertilizers' usage. The use of fertilizers and water can be optimized using smart systems [6]. Various technologies, including remote sensing to geotechnical engineering (satellite communication and image processing), are available [7 - 9]. Crop disease is a serious major issue that directly affects crop production. Proper monitoring of crop disease is required. Farmers use pesticides to overcome plant diseases, but they are not much aware of crop pesticide requirements. Pesticides may either affect the yield or be injurious to consumers' health. A smart system that can estimate the number of pesticides using training data is proposed [10 - 12].

There are many steps in cultivation, from the beginning of the crop until its maturity. Among them are crop planting, weeding, watering from time to time, and spraying medicine to avoid diseases. Monitoring of soil moisture using sensors and optimizing water and fertilizer supply is required [13]. In order to supply high-value AI applications in the aforementioned industry, agricultural robots are constructed. The development of smart irrigation technology enables farmers to boost productivity without using a lot of labour by monitoring soil temperature, fertilizers' content, water level, and weather forecasts. Turning the irrigator pump ON/OFF causes the microcontroller to initiate the actuation [14]. The primary goal of the moisture estimation method is to provide a real-time system for measuring soil moisture. The technique is based on the relationship between two variables, specifically the sound speed and the rate at which soils absorb water. According to this experiment, depending on the kind of soil, the sound speed decreases as the moisture content increases [15].

A smart system combines sensing, control, activation, analysis, and decision-making based on different automation operations [16]. The automation processes rely on a variety of novel technologies, networking capabilities, sensing capabilities, and intelligence calculations. Therefore, with the aid of a smart system, we can efficiently utilize water and boost yields. Agriculture based on cutting-edge technologies and intelligent irrigation systems uses less freshwater [17]. Artificial intelligence is used in agriculture to enhance irrigation and pesticides and reduce chemical usage through image processing [18]. A hybrid approach combining AI neural networks, fuzzy logic, and genetic optimization techniques is developed to forecast each agronomist's daily irrigation depth [19, 20]. Through effective and dependable techniques like wireless sensor networking, Global System for Mobile Communication, and short message service technologies. Farmers can avail of this service easily on mobile phones. This system offers this capability that undoubtedly helps farmers increase their yields and, on a macro level, aids agricultural and national economic expansion [21, 22]. The successful forecasting and management of grain production will benefit from the multidisciplinary approach of merging computer science with agriculture [23].

The real-time feedback control system used by the microcontroller drip irrigation system effectively monitors and manages all system operations [24]. The solar-based irrigation system incorporated with moisture sensors, a microcontroller, and a global system for mobile communication module is proposed by Sawant *et. al.* [25]. WSN, and ZigBee-like techniques are discussed to achieve better performance, and optimized and economical systems [26].

As a result of recent technological advancements, it is now important to use smart irrigation systems to boost the agricultural crop's productivity level. The present irrigation system optimizes water supply to the land in respect of various parameters such as soil moisture, land slope, and climate condition. This system is empowered with a solar panel to reduce cost and grid dependency. AI and IoT techniques facilitate the manual operation of irrigation systems. Manual intervention of farmers is almost eliminated for supplying water to the field in all weather conditions. The payback period of the entire cost of the system is also calculated to check the feasibility of the smart irrigation system.

METHODOLOGY

The solar pumping module and the intelligent irrigation module constitute the present artificial intelligence-integrated irrigation system. A solar panel is placed near the pump set. The system includes a water level sensor that will signal whether there is water in the tank. A microcontroller is configured with the automatic threshold values of soil moisture and water level to advance a software application. This software application also considered the prediction of the amount of rainfall and the soil moisture content. Beneath the earth, where plant roots may reach them, sensors for soil temperature and moisture are positioned. The Photovoltaic solar panel is also incorporated with a microcontroller through a relay switch to the operated pump. The methodology of the present smart irrigation system is described in Fig. (**1**). Training data such as output from soil moisture, temperature sensors, and weather data are fed to the microcontroller to train the system. The historical past data is also provided to the microcontroller. On the basis of these input feuded data, the microcontroller can estimate and optimize the crop requirement.

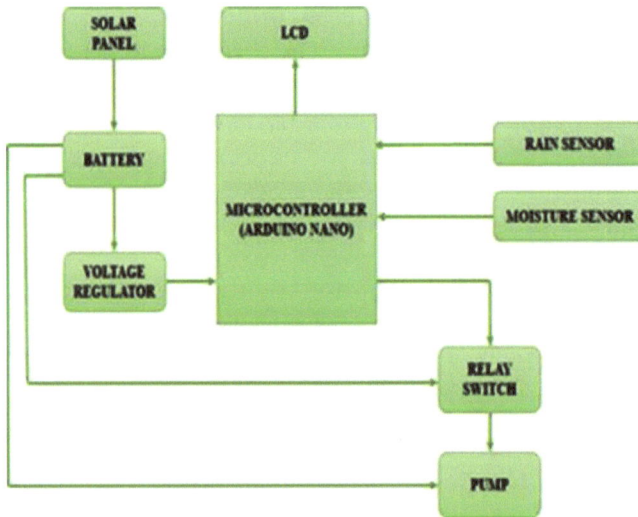

Fig. (1). Block diagram of the main part of the auto solar irrigation system.

SYSTEM AND COMPONENTS

The following hardware component is assembled for the implementation of the present irrigation system. Fabrication of the prototype is carried out in the Mechatronics and IoT lab of the Mechanical Engineering Lab and Center of Alternate & Renewable Energy (CARE) of Rajkiya Engineering College,

Azamgarh. The fabrication process consists of electronic fabrication, soldering, wood cutting, and drilling. Description of the hard component is as follows:

Solar Panel

A photovoltaic (PV) module is an interconnection set-up of solar units (usually 6×10 type) which provide DC supply. Each module is specified by its watt rating under standard operating conditions and generally lies in a few hundred Watts (max limit 500 Watts). The size of a module for a certain rated output depends on the module's efficiency. The efficiency of solar cells is up to 22 percent (polycrystalline material) and 27 percent (monocrystalline material). PV panel systems make up an array, and a PV panel is a grouping of PV modules.

Lead Acid Battery

The oldest and most developed technology is found in lead-acid batteries. Two electrodes are immersed in an electrolyte of sulfuric acid to form the lead-acid battery. The negative electrode is linked to a grid of metallic lead, whereas the positive electrode is formed of metallic lead oxide grains. In Valve-regulated lead–acid (VRLA), batteries are sometimes known as "maintenance-free" systems since no replacement of water lost during electrolysis is necessary [27].

Stepper Motor

A stepper motor is a BLDC motor that, unlike many other common types of electric motors, doesn't merely spin endlessly until the DC voltage going to it is turned off. Stepper motors are an alternative form of digital input-output device for precise opening and ending. They are designed to be quickly turned on and off, allowing the motor to turn through a small fraction of a revolution at a time. These discrete, planned phases are referred to as "steps".

In order to move through certain rotational degrees or angles, stepper motors are made to divide a single full revolution into a number of significantly smaller part rotations. As a result, mechanical parts that need a high level of accuracy may be moved with a stepper motor's precision down to the micron. Stepper motors are frequently controlled digitally and serve as essential elements of an open-loop motion control positioning system. Their capacity to impose far more precisely specified rotational locations, speeds, and torques makes them excellent for activities requiring highly stringent movement control, where they are most frequently utilised in holding or positioning applications. Four high-torque DC motors are used to power the robot. The microprocessor and motor driver (L293D) are interfaced to control the robot.

5V Power Supply Using 7805 Voltage Regulator

The design of the power supply includes the LM7805 since the microcontroller (ATMEGA-328) needs a tightly controlled, continuous 5V supply. A regulated power supply includes a filter and Zener diode.

Micro-Controller

An onboard power supply, USB port, and Atmel microcontroller chip are all included on a microcontroller board [28, 29]. By offering a standard board that can be programmed and linked to the system without the need for complex PCB design and execution, it makes the process of designing any control system simpler based on the ATmega328 as shown in Table **1**. The ATMEGA 328 microcontroller, which sends the interrupt signal to the motor, controls the entire system. Temperature sensors and humidity sensors are associated with the internal ports of the microcontroller through a comparator. When there is a change in the environment's temperature and humidity, these sensors detect it and send interrupt data to the microcontroller, which starts the motor. A buzzer is also used to signal that the pump is turned on [30, 31].

Table 1. Specifications of Arduino Nano [30, 31].

Microcontroller	Atmel ATmega168 or ATmega328
Operating Voltage (logic level)	5 V
Input Voltage	7-12 V
Digital I/O Pins	14 (of which 6 provide PWM output)
Analog Input Pins	8
DC Current per I/O Pin	40 Ma
Flash Memory	16 Kb (ATmega168) or 32 Kb (ATmega328) of which 2 Kb used by boot loader
SRAM	1 Kb (ATmega168) or 2 Kb (ATmega328)
EEPROM	512 bytes (ATmega168) or 1 Kb (ATmega328)
Clock Speed	16 MHz

Motor Driver L293D

For operating stepper Motors, the L293D Motor Driver Module is an ideal medium power motor driver. It can turn on and off four DC motors or regulate the direction and speed of two DC motors. These drivers considerably simplify and improve the convenience of using microcontrollers to operate motors, relays, and other devices. The L293D is designed to provide up to 600 mA of bidirectional driving current at voltages between 4.5 V and 36 V. This driver is best suited for the bidirectional movement of motors.

Moisture Sensor

This sensor may be used to measure the soil's moisture; when the soil lacks water, the module's output is high; otherwise, the output is low [32]. The flowering plant or any other plant that requires automated watering may be watered automatically with this sensor. The module has three output modes: easy digital output, accurate analogue output, and precise serial output.

LCD 16×2

A liquid-crystal display (LCD) is an electronic display panel. It employs liquid crystals' ability to modulate light. Each pixel of a liquid crystal display, which is organized into a rectangular network, has a backlight that illuminates it. A 16x2 LCD features two horizontal lines, each of which contains a space for 16 characters to be shown.

Relay Switch Circuit

Relays are electromechanical devices that employ an electromagnet to move a pair of moveable contacts from an open state to a closed position. Relays have the benefit of being able to regulate AC circuits, motors, heaters, and other devices that can consume a lot more electrical power while only requiring a tiny amount of electricity to run the relay coil. The assembled view of the project having all components mentioned above is depicted in Fig. (**2**).

Fig. (2). An Artificial Intelligence Integrated Irrigation System.

WORKING

The solar pumping module and the autonomous irrigation module make up most of the proposed irrigation system. A solar panel with the necessary specifications is positioned next to the pump set in the solar pumping module. The battery is charged through the solar panel. The water pump is submerged into the source of water and operated by the battery power. The water is then circulated into the tank, where it is initially stored before being discharged into the field. A soil moisture detecting circuit in an autonomous irrigation module electronically controls the water output valve of the tank. The crop has been farmed as the sensors are placed all over the area. The sensor transforms the equivalent voltage of the soil's moisture level. A sensor circuit with a reference voltage is supplied with an equivalent voltage. The farmer can change this reference voltage to set various moisture levels for various crops.

The difference between these two voltages directly relates to the amount of water required for soil. A stepper motor whose rotational angle is inversely correlated with the voltage difference was given a control signal. The cross-sectional area of the valve that has to be opened to regulate water flow is controlled by the stepper motor in turn. As a result, the flow of water is proportional to the difference in moisture content.

Moisture Estimation

The sensors, which essentially assess the moisture content, figure out how much moisture is in the soil. Herein we implement an intelligent irrigation system. The method of sensing is the dielectric method and neutron moderation method. In a dielectric method, sensors measure moisture in real-time control based on the dielectric properties. The estimation of the soil's dielectric consistency is based on how each of these components is typically contributed because the soil is made up of several constituents, including minerals, air, and water. The proximity of moisture in the soil mostly influences the projected value of permittivity. Compared to the other soil sections, where the estimated value of this constant is significantly smaller (Kaw = 81), water has a much higher estimated dielectric value. The equation of Topp *et al.* where; dielectric constant (Ka_b) and volumetric soil moisture (VWC) [33],

$$VWC = -5{:}3 \times 10^{-2} + 2{:}29 \times 10^{-2}\, Ka_b - 5.5 \times 10^{-4}\, Ka_{(b)}{}^2 + 4.3 \times 10^{-6}\, Ka_{(b)}{}^3 \dots \quad (1)$$

The above equation can be used to determine the dielectric constant (Kab).

$$K_{ab} = (c/v)\, 2 = ((ct)/2L)/2 \dots \quad (2)$$

Another method of moisture determination is neutron moderation. In this method, fast neutrons are fired from a dynamic radio source that is breaking down, such as 241Am/9Be [34], and when they collide with protons or H+, they swiftly decrease, creating a thermalized neutron cluster. V is the propagation velocity of an element.

Where c is the speed of electromagnetic waves in a vacuum and L is the length of the TL in the soil (in m or ft), the transmission time (t in a flash).

This system determines the crop's irrigation needs by weighing in the required water level content. In conventional applications, the farmer or an agricultural expert makes these selections based on their experience. Data is received from a variety of sources, including meteorological statistics, crop and soil parameters, information gathered from sensors installed in the soil, and moisture-influencing factors like evapotranspiration [35]. Our self-sustained irrigation system is created on the basis of this concept. We compile past data, which comprises actions suggested by a farming expert regarding the amount of water needed by a crop based on the facts at hand and the state of the crop at the time. Artificial Intelligence is used to assess the data collected in order to estimate the crop's real-time water needs.

The agricultural expert's decision reports and past data will be used to develop the AI system while also taking into account the different real-time elements discussed earlier and the known water level requirements of the specific crop. The AI model aims to be similar to the judgments made by an agronomic, which are utilized as the basis for evaluation. To obtain the highest performance with the least amount of computing needed, few predictive algorithms [36] are used as proposed in the literature. A schematic diagram of the AI system is shown in Fig. (3).

The amount of water that has to be irrigated to the crop depends on the water level (moisture content) required to make up for the moisture loss from the field due to evapotranspiration. Therefore, one of the key determining factors in determining the crop's irrigation needs is determining the evapotranspiration of a crop area. Weather station data or data from weather sensors is used to calculate the reference crop evapotranspiration. (ET0) daily is calculated using the FAO Penman-Monteith formula.

$$E\ Tc = K.\ E\ To \dots \tag{3}$$

Where, Kc factor serves as an accumulation of physiological and physical variations between crops and (E. To) is defined as crop reference evapotranspiration.

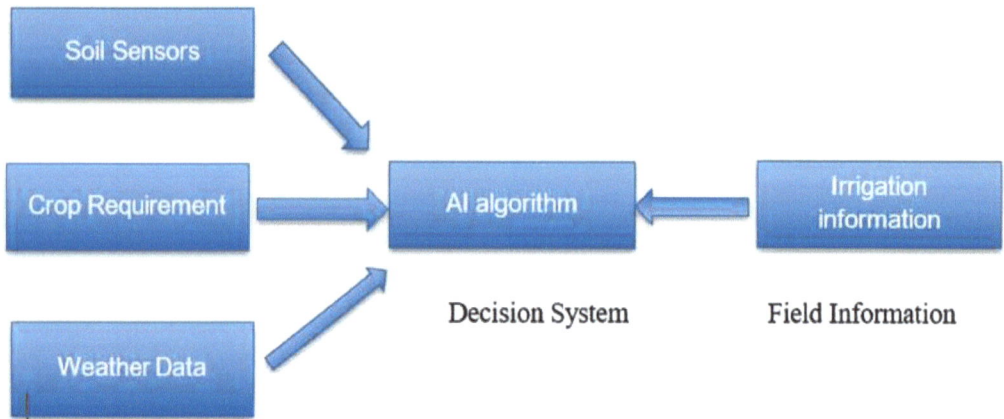

Training Information

Fig. (3). Moisture Estimation AI system [37].

PAYBACK PERIOD CALCULATION

It is the amount of time after an investment has been made that the equipment, facility, or other investment has generated enough net income to pay the cost of the investment. Land area: 1 acre (4046.86 m^2), Rating of Pump: Power = 900watt, Discharge = 81000 LPD, Depth of irrigation: 0.07 meter are considered.

The total discharge evaluated for the above parameters is Q = 283.2767 m^3.

The time taken by the pump for irrigation of land area is 83.933 hours for irrigation.

The total energy consumed by the pump for irrigation = 75.54 kWh.

The cost for 1 unit of electricity consumption we pay is = Rs. 10.00.

Total Money spent is = (75.54×10) = Rs. 755 (it was money spent on irrigation at one time).

Let us suppose that there is five times the requirement of irrigation for a crop in a month, so, Total money spent is = 755×5= Rs 3775.

Suppose that irrigation happens for six months in a year,

Total cost paid is = 3775×6 = Rs 22650.

Cost of smart irrigation system.

As per review and market conditions, the average cost of a 1-watt set-up is Rs 25.

The cost of the 900-Watt system, including installation is = Rs 25000.

The time of payback cost is = 1.1 years per acre irrigation.

The proposed AI solar-empowered irrigation system will be cost-free for 1-acre irrigation after 1.1 years of working.

CONCLUSION

The worldwide agriculture sector is facing water and energy shortages in the future, making smart irrigation systems a critical study topic. Therefore, a key job in a smart irrigation system is to significantly use water, reduce dependency on grid energy, avoid expenses, and boost crop yield. This solar-powered smart irrigation system increases the reliability of farmers in terms of dependency on power cuts, fuel availability, liquid money, and timing of water supply. Due to its self-starting nature, the system requires less maintenance and care. The government and farmers are likely to benefit from the proposed system's implementation in several ways. This chapter discusses all the required parts of smart irrigation systems and their coordination with artificial intelligence. Crop yield production is a function of soil moisture, crop evaporation, as well as other diverse factors. Methods of moisture estimation and crop evaporation are also discussed. The present smart irrigation is trained with all historical data and ground data to optimize water supply to the crop using a smart system. This smart system also reduces farmers' manual effort in adverse weather conditions and saves their time. The payback period of this prototype is calculated as 1.1 years, reflecting the cost-effectiveness of the system.

ACKNOWLEDGEMENTS

The authors are grateful to the Centre of Renewable and Alternative Energy (CARE) Lab, Mechatronics and IoT labs of the Mechanical Engineering Department, Rajkiya Engineering College, Azamgarh UP, for providing lab facilities.

REFERENCES

[1]   S. Siebert, J. Burke, J.M. Faures, K. Frenken, J. Hoogeveen, P. Döll, and F.T. Portmann, "Groundwater use for irrigation – a global inventory", *Hydrol. Earth Syst. Sci.,* vol. 14, no. 10, pp. 1863-1880, 2010.
[http://dx.doi.org/10.5194/hess-14-1863-2010]

[2]   Available at:https://www.worldbank.org/en/topic/water-in-agriculture

[3]   *Agricultural irrigated land (% of total agricultural land).* The World Bank, 2013.

[4]   Economic Times, *How to solve the problems of India's rain-dependent on agricultural land.*

[5]   L.A. Zadeh, "Fuzzy sets", *Inf. Control,* vol. 8, no. 3, pp. 338-353, 1965.
        [http://dx.doi.org/10.1016/S0019-9958(65)90241-X]

[6]   K. Bhagyalaxmi, K. K. Jagtap, N. S. Nikam, K. K. Nikam, and S. S. Sutar, "Agricultural robot"
        (Irrigation ystem, weeding, monitoring of field, disease detection)", *International Journal of
        Innovative Research in Computer and Communication Engineering.,* vol. 4, no. 3, pp. 4403-4409,
        2016.

[7]   P. S., D. P. Mahato, and N. T. D. Linh, *Distributed Artificial Intelligence.,* S.P. Yadav, D. P. Mahato,
        N. T. D. Linh, Eds., CRC Press.: Boca Raton, 2020.
        [http://dx.doi.org/10.1201/9781003038467]

[8]   S.A. Ackerman, "Remote sensing aerosols using satellite infrared observations", *J. Geophys. Res.,* vol.
        102, no. D14, pp. 17069-17079, 1997.
        [http://dx.doi.org/10.1029/96JD03066]

[9]   S.P. Yadav, and S. Yadav, "Fusion of Medical Images in Wavelet Domain: A Discrete Mathematical
        Model", In: *In Ingeniería Solidaria.* vol. 14. Universidad Cooperativa de Colombia: UCC., 2018, no.
        25, pp. 1-11.
        [http://dx.doi.org/10.16925/.v14i0.2236]

[10]   S. Arivazhagan, "Newlin. R. Shebiah R., S. Ananthi, S. Vishnu, Varthini, "Detection of unhealthy
        region of plant leaves and classification of plant leaf diseases using texture features"", *Agric. Eng. Int.
        CIGR J.,* vol. 15, no. 1, pp. 211-217, 2013.

[11]   K.P. Ferentinos, "Deep learning models for plant disease detection and diagnosis", *Comput. Electron.
        Agric.,* vol. 145, pp. 311-318, 2018.
        [http://dx.doi.org/10.1016/j.compag.2018.01.009]

[12]   T. Hemalatha, and B. Sujatha, "Sensor based autonomous field monitoring agriculture robot providing
        data acquisition & wireless transmission", *International Journal of Innovative Research in Computer
        and Communication Engineering,* vol. 3, no. 8, pp. 7651-7657, 2015.

[13]   Yunseop Kim, R.G. Evans, and W.M. Iversen, "Remote sensing and control of an irrigation system
        using a distributed wireless sensor network", *IEEE Trans. Instrum. Meas.,* vol. 57, no. 7, pp. 1379-
        1387, 2008.
        [http://dx.doi.org/10.1109/TIM.2008.917198]

[14]   *Artificial Intelligence in Agriculture.* vol. 4. , 2020, pp. 58-73.

[15]   V. Vashisht, A.K. Pandey, and S.P. Yadav, "Speech Recognition using Machine Learning", In: *In IEIE
        Transactions on Smart Processing & Computing.* vol. 10. The Institute of Electronics Engineers of
        Korea., 2021, no. 3, pp. 233-239.
        [http://dx.doi.org/10.5573/IEIESPC.2021.10.3.233]

[16]   G. Akhras, "Smart Materials and Smart Systems for the Future", *Military Journal,* 2000.

[17]   J. Gubbi, R. Buyya, S. Marusic, and M. Palaniswami, "Internet of Things (IoT): A vision, architectural
        elements, and future directions", *Future Gener. Comput. Syst.,* vol. 29, no. 7, pp. 1645-1660, 2013.
        [http://dx.doi.org/10.1016/j.future.2013.01.010]

[18]   T. Talaviya, D. Shah, N. Patel, H. Yagnik, and M. Shah, "Implementation of artificial intelligence in
        agriculture for optimisation of irrigation and application of pesticides and herbicides", *Artificial
        Intelligence in Agriculture,* vol. 4, pp. 58-73, 2020.
        [http://dx.doi.org/10.1016/j.aiia.2020.04.002]

[19]   R. González Perea, E. Camacho Poyato, P. Montesinos, and J.A. Rodríguez Díaz, "Prediction of
        applied irrigation depths at farm level using artificial intelligence techniques", *Agric. Water Manage.,*
        vol. 206, pp. 229-240, 2018.
        [http://dx.doi.org/10.1016/j.agwat.2018.05.019]

[20]   J. Angelin Blessy, "Smart Irrigation System Techniques using Artificial Intelligence and IoT", *Third International Conference on Intelligent Communication Technologies and Virtual Mobile Networks (ICICV,* pp. 1355-1359, 2021.
[http://dx.doi.org/10.1109/ICICV50876.2021.9388444]

[21]   S. Kalyan, K. Gopinath, T. Govindaraju, N. Devika, and V. Suthanthira, "GSM based Automated IrrigationControl using Raingun Irrigation System", *Int. J. Adv. Res. Comput. Commun. Eng.,* vol. 3, no. 2, 2011.

[22]   S. Maleekh, S. Dhanne, S. Kedare, and S. S. Dhanne, "Modern Solar Powered Irrigation System by Using ARM", *International Journal of Reseach in Engineering and Technology.,* vol. 3, no. 3, 2013.

[23]   S. Yethiraj, R. Harishankar, S. Kumar, K.P Sudharsan, U. Vignesh, and T. Viveknath, "Solar Powered Smart Irrigation System", *dvance in Electronic and Electrical Engineering.,* vol. 4, no. 4, 2012.

[24]   D. Prathyusha, N. Lakshmi, and K.S. Gomathi, "Smart Irrigation System Automation Monitoring and Controlling of Water Pump by using Photovoltaic Energy", *SSRG International Journal of Electronics and Communication Engineering.,* vol. 2, no. 11, 2012.

[25]   S. Kumar, R. S. Sawant, S. Gubre, S. Pillai, and M. Jain, "Solar Panel Based Automatic Plant Irrigation System", *International Journal of Innovative Science, Engineering and Technology.,* vol. 2, no. 3, 2013.

[26]   H.P. Patel, "Garg", In: *Advances in solar energy technology* vol. 3. Reidel Publishing: Boston, M.A, 2012.

[27]   M. Skyllas-Kazacos, "10 - Electro-chemical energy storage technologies for wind energy systems", *Stand-Alone and Hybrid Wind Energy systems.,* pp. 323-365, 2010.

[28]   *The 8051 Microcontroller and Embedded Systems Using Assembly and C - Muhammad Ali Mazidi.* Easy Engineering, 2006.

[29]   B.K. Chate, "Smart Irrigation System using Raspberry Pi", *International Research Journal of Engineering and Technology (IRJET).,* vol. 3, no. 5, p. 2395-0072, 2016.

[30]   K. Liakos, P. Busato, D. Moshou, S. Pearson, and D. Bochtis, "Machine Learning in Agriculture: A Review", *Sensors (Basel),* vol. 18, no. 8, p. 2674, 2018.
[http://dx.doi.org/10.3390/s18082674] [PMID: 30110960]

[31]   B. D. Kumar, P. Srivastava, R. Agrawal, and V. Tiwari, "Microcontroller based automatic plant irrigation system", *International Research Journal of Engineering and Technology (IRJET).,* vol. 4, no. 5, 2017.

[32]   *Programming in AVR tutorials.* iBOT: Robotics club of IIT Madras, 2006.

[33]   M.F. Gebregiorgis, and M.J. Savage, "Determination of the timing and amount of irrigation of winter cover crops with the use of dielectric constant and capacitance soil water content profile methods", *S. Afr. J. Plant Soil,* vol. 23, no. 3, pp. 145-151, 2006.
[http://dx.doi.org/10.1080/02571862.2006.10634746]

[34]   I.F. Long, and B.K. French, "Measurement Of Soil Moisture In The Field By Neutron Moderation", *Eur. J. Soil Sci.,* vol. 18, no. 1, 1967.

[35]   R.G. Allen, *An Update for Calculation of Evapotranspiration Reference.* vol. 43. ICID Bulletin, 1994, p. 2.

[36]   S. Wold, A. Ruhe, H. Wold, and I. W. Dunn, *The Collinearity Problem in Linear Regression. The Partial Least Squares (PLS) Approach to Generalized Inverses,* 1984.
[http://dx.doi.org/10.1137/0905052]

[37]   S. Choudhary, G. Vipul, A. Singh, and S. Agrawal, "Autonomous Crop Irrigation System using Artificial Intelligence", *Int. J. Eng. Adv. Technol.,* vol. 8, no. 5S, 2019.

CHAPTER 9

Leveraging AI for Smart Cities in India

Manisha Singh[1,*]

[1] *Economics and Strategy, G.L. Bajaj Institute of Management and Research, Greater Noida, India*

Abstract: With the fast spread of connectivity *via* 5G and IoT (Internet of Things), the Smart City Artificial Intelligence (AI) software industry is expected to reach a massive value of $ 4.9 billion by 2025 globally [1]. In India, the AI market is slated to reach $ 7.8 billion by 2025 at a CAGR of 20.2% as per an International Data Corporation (IDC) report [2]. This chapter explains how AI can be used in the ambitious Smart Cities Mission (SCM) announced by the Government of India in 2015 [3]. Beginning with the conceptual understanding of the SCM, the chapter introduces AI as a useful aid to urban planning thereby creating a safer and sustainable future for its citizens. Applications of AI in Smart cities are then discussed followed by a brief discussion on the prevailing best practices. Challenges in creating AI-enabled smart cities in India are outlined followed by the conclusion which chalks out the road ahead for AI-enabled smart cities in India.

Keywords: Artificial Intelligence, Smart Cities, Urbanization.

INTRODUCTION

In 2020, 56% of the world's population – 4.4 billion inhabitants – lived in cities. This trend is expected to continue. By 2050, with the size of the urban population will get double, about 7 of 10 people in the world will live in cities [4]. Though urbanization is potentially growth stimulating for any economy, unplanned urbanization can lead to congestion, poor quality of life, over-burdened infrastructure, pollution, *etc*.

To mitigate these damaging spill-overs, the Government of India conceptualized the Smart Cities Mission (SCM) with the aim of driving economic growth and improving the quality of life of people by encouraging local area development and using technology that can lead to smart outcomes.

With a proposed Investment of ₹ 2,05,018 crores, 7,831 projects were tendered at ₹ 1,91,337 crores, 4,161 projects have been completed using up ₹ 68,155 crores, as of August 11, 2022.

---

[*] **Corresponding author Manisha Singh:** Economics and Strategy, G.L. Bajaj Institute of Management and Research, Greater Noida, India; Tel: 8527820126; E-mail: manishasing@gmail.com

The Smart Cities Mission was launched on June 25th, 2015 as a joint effort of the Ministry of Housing and Urban Affairs (MoHUA) and all state and union territories. It had aimed to be completed by 2019-20, but has been extended to June 2023 as per the latest information available. (Fig. **1**).

Fig. (1). SCM -Plan of Investment (Source: https://smartcities.gov.in) [3].

Hundred cities and towns across States and UTs of India have been selected under this mission which house more than one-third of the country's population. The SCM ensures that people are able to find opportunities in a sustainable and enabling environment. In other words, according to MoHUA, "smart cities are cities that work for the people."

IMPLEMENTATION OF SCM

The implementation of the mission at city level is proposed to be done by a **Special Purpose Vehicle (SPV)** that will plan, appraise, approve, release funds, implement, manage, operate, monitor and evaluate the Smart City development projects. Like any other developer, the SPV would develop a site, generate revenues for itself and exit after completion. This may take the form of joint ventures, subsidiaries, public-private partnership (PPP), turnkey contracts, *etc*.

Registered under the Companies Act, 2013 at the city-level, it would have an equity shareholding of 50% each distributed between the ULB (Urban Local Body) and State/UT. Government funding can take the shape of grants with the MoUD (Ministry of Urban Development) specifying the conditions thereof. (Fig. **2**).

Fig. (2). Smart Cities in India (Source: www.mapsofindia.com) [14].

It is proposed to give full flexibility to SPV to implement and manage the Smart City project. This may be seen in their decision of appointing PMCs (Project Management Consultants) and using frameworks developed by MoUD for the projects.

ARTIFICIAL INTELLIGENCE

As per Britannica, 'artificial intelligence (AI) is defined as the ability of a digital computer or computer-controlled robot to perform tasks commonly associated with intelligent beings.' According to IBM, 'Artificial intelligence leverages computers and machines to mimic the problem-solving and decision-making capabilities of the human mind'.

In 1950, Alen Turing, often regarded as the father of Computer Science, in his pioneering work, 'Computing machinery and Intelligence', posed questions like 'can machines think'?

Simply stated, Artificial Intelligence combines computer science and complex datasets for problem-solving (Fig. **3**). As the case with any new and emerging technology, AI has created a lot of curiosity and excitement around it. As per Gartner's Hype Cycle (Figs. **4** & **5**), product innovations follow typical progressions of innovation, starting with overenthusiasm through a time of disillusionment to a gradual understanding of the relevance of the innovation and place in the market [10] (Fig. **4**).This seems to be true for self-driving cars and personal assistants. In an MIT lecture in 2019, Lex Fridman observed that we may be at the peak of inflated expectations travelling down the trough of disillusionment [5].

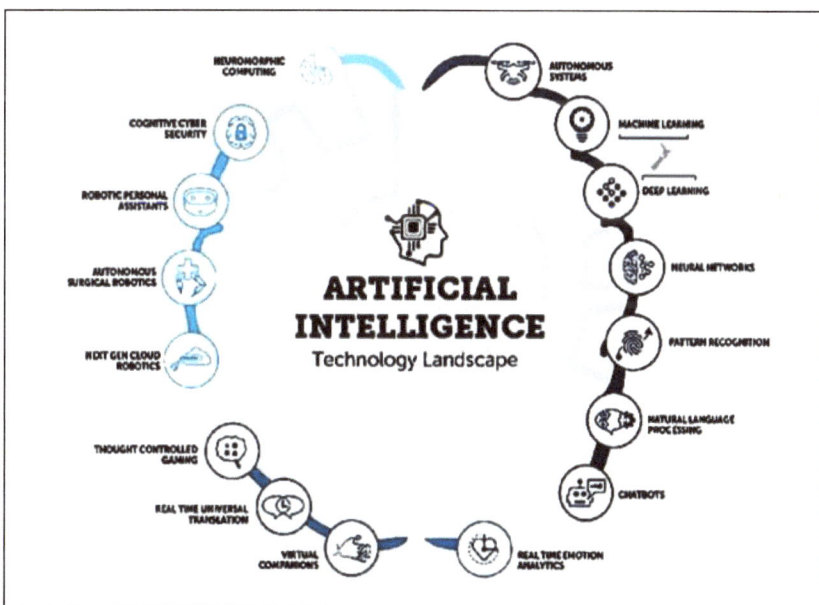

Fig. (3). Artificial Intelligence Technology Landscape (Source: AI Time Journal) [6].

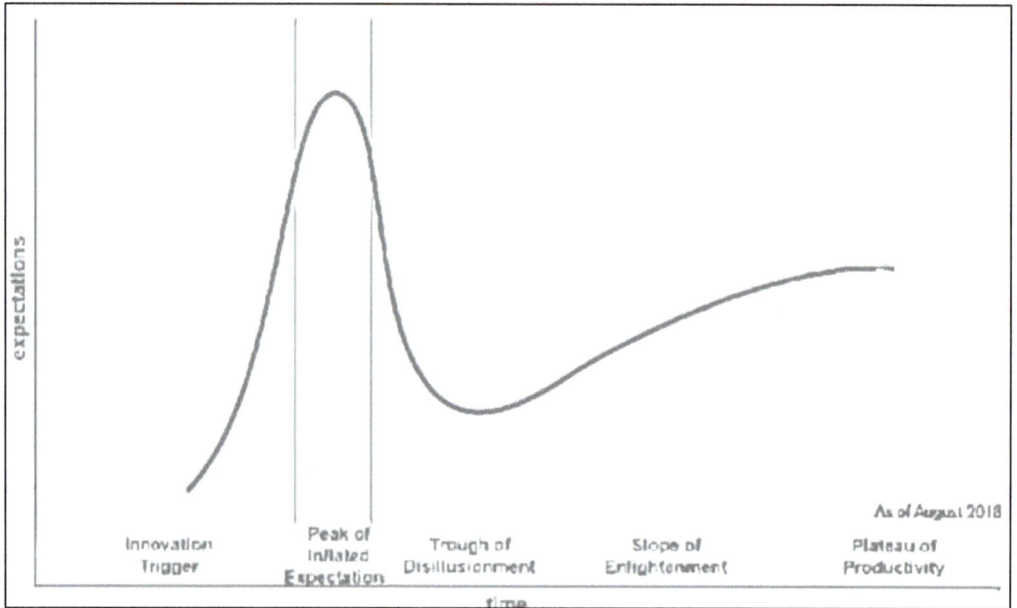

Fig. (4). The Hype Cycle (Source: Gartner Aug, 2018) [10].

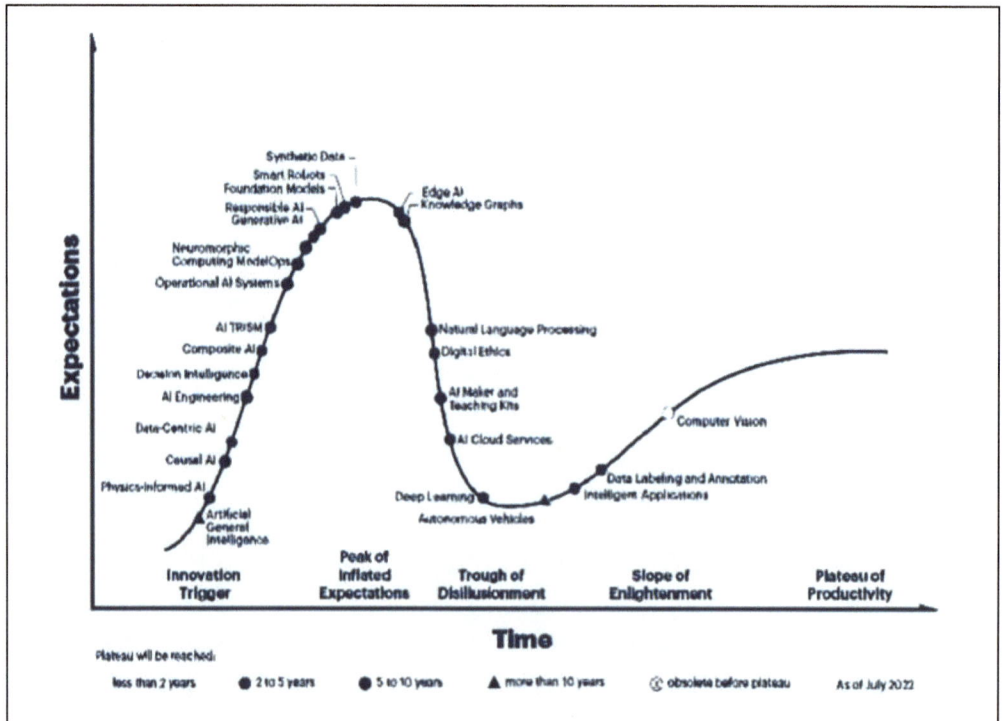

Fig. (5). Hype Cycle for Artificial Intelligence (Source: Gartner, 2022) [10].

WEAK AI AND STRONG AI

Weak or more aptly, Narrow AI is what we see around us most of the times-Amazon's Alexa, Apple's Siri, self-driving cars. This AI focuses on performing one type of task. Strong AI comprises both Artificial General Intelligence (AGI) which is comparable with human capabilities and Artificial Super Intelligence (ASI) which is more capable than a human. Currently, Strong AI is entirely theoretical, but researchers are exploring it.

LEVERAGING AI FOR SMART CITIES

It is estimated that more than 30% of smart city applications would be based on AI by 2025, including urban transportation solutions, substantially contributing to the sustainability, resilience, vitality and welfare of urbanites [6, 12].While AI finds easy application in many areas and sectors including speech recognition, customer service, computer vision, recommendation engines, automated stock trading, we would be focussing on leveraging AI for smart cities. As already discussed, Government of India envisages using AI to optimize its Smart Cities Mission as seen below in (Fig. **6**).

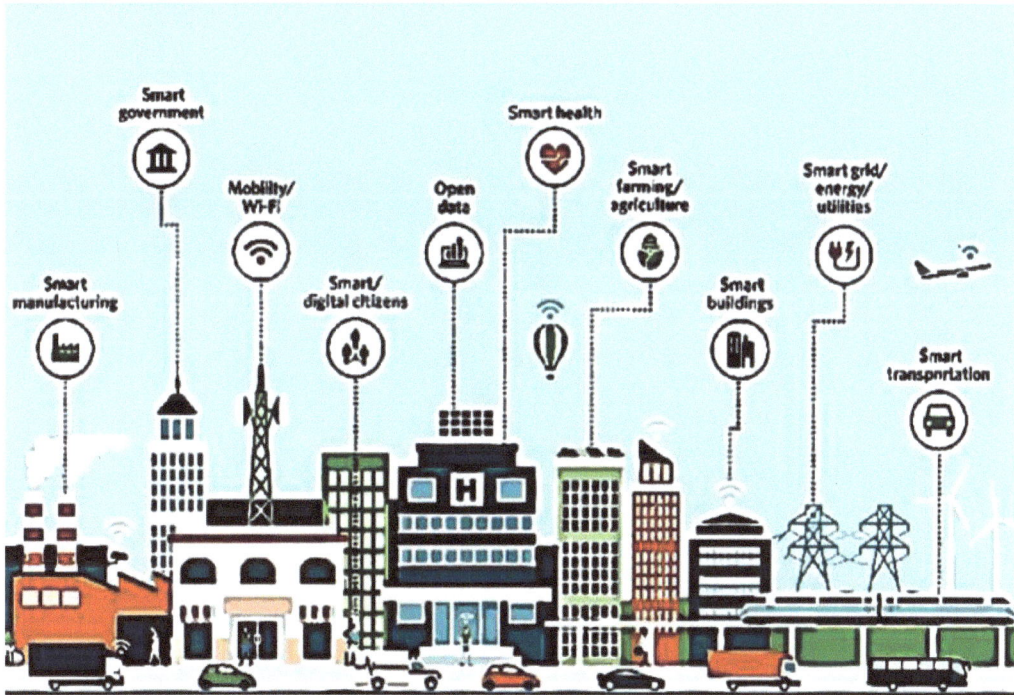

Fig. (6). AI applications in Smart City (Source: Traballesi *et al.* 2019) [13].

Google trend shows the increasing interest in terms 'smart cities' and 'AI' since 2014 as below in (Fig. **7**):

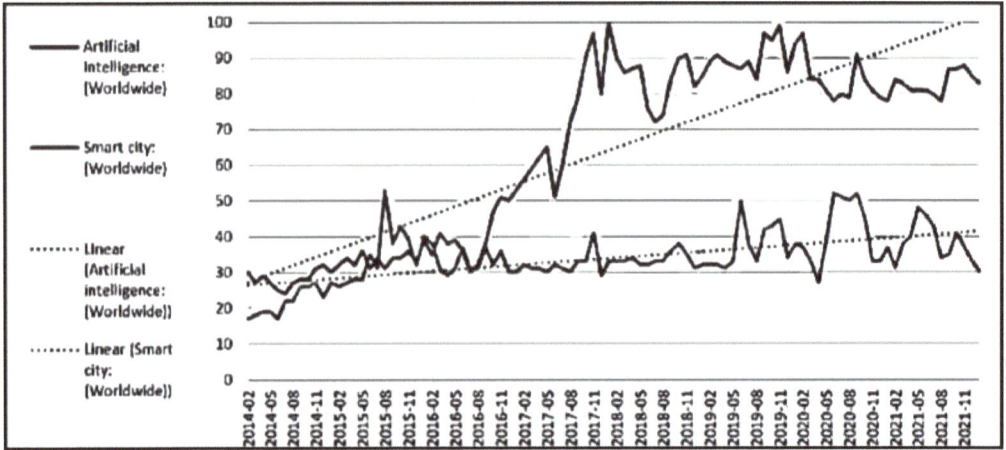

Fig. (7). The popularity of the keywords "*Smart City*" and "*Artificial Intelligence*" since 2014, *(Source:* H.M.K.K.M.B. Herath, Mamta Mittal, 2022) [11].

The scope of AI in smart city can be known from following (Fig. **8**):

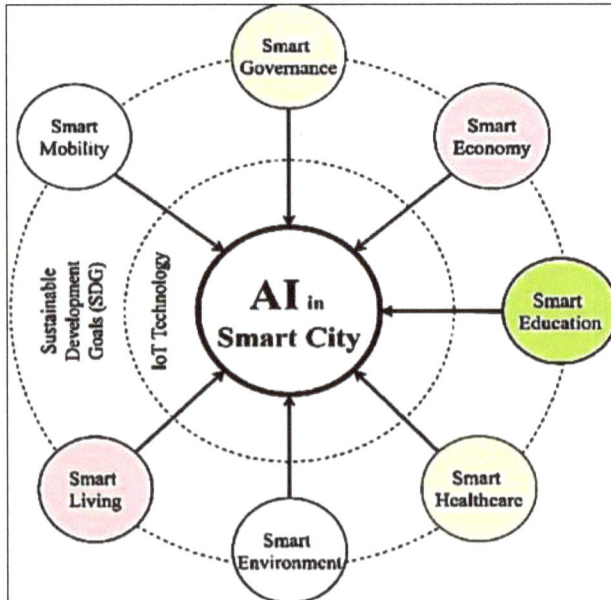

Fig. (8). Scope of AI in Smart City (*Source:* H.M.K.K.M.B. Herath, Mamta Mittal, 2022) [11].

The Indian Government can use AI in the following areas to make its cities smart:

Security & Surveillance

An AI-powered surveillance system can undoubtedly be the solution to a robust and full proof mechanism of locating, recording and even predicting criminal activities by connecting surveillance cameras to an AI network. By using drone enabled surveillance cameras, quick alerts can be created if any suspicious activities are detected.

Traffic Management and Vehicle Parking

Road safety can be enhanced using AI powered smart traffic signals which re-adjust their timing according to the flow of traffic, which can reduce traffic as well as result in greater fuel efficiency. AI sensors in parking lots can indicate whether parking spot is available or not.

Waste Collection and Disposal Management

Cities with high density of population generate large amounts of waste and their efficient management is crucial for sustainable quality of life. AI sensors in garbage cans can indicate when the can is full and ready for collection, saving fuel of the collecting authority and charting new smart routes for optimizing waste collection. Also, AI-enabled machines can be used to segregate the different types of waste, for example: recyclable or non-recyclable.

Energy Management

Energy consumption can be optimized using AI. Smart streetlights can detect where there is more traffic and automatically dim the lights in areas with low traffic. AI can be used to analyse and monitor citizen's use of energy so that renewable sources may be brought into use promptly.

Environment and Pollution

AI can be effectively used to see the effect of climate change, global warming and pollution on the city through predictive tools that analyse environmental data and create self-configuring weather forecasts.

Planning Management and City Administration

With the help of an AI network, vast information about the city in terms of inter-city population densities and traffic flows can aid the Government in the optimum utilization of resources as per the need. AI based technologies can be used to analyse road imagery to assess when repairs are required.

BEST PRACTICES

An integration of Machine Learning, AI, IoT and ICT can create efficient smart cities with the help of the following best practices:

Connectivity

Processing large data requires smooth connectivity and a plurality of technologies like 4G, 5G, Wireless, LAN, Bluetooth would be desirable as no single technology provides all solutions.

Scalability

Scalability refers to the extent to which the size and power of a system can accommodate changes in storage and workflow demands. In case of smart cities, its infrastructure needs to be capable of taking care of future needs.

Security

Trust between the citizen and the Government is paramount for any such initiative and there is a need for a clear policy protecting citizen's private data.

Partnership

The choice of the technology partner must be based on considerations like innovation capacity, investment capacity and experience given the sensitivity, complexity and public-private-partnerships which define the ecosystem of smart cities.

Citizen Engagement

As citizens are the most important stakeholders in a smart city, their active and meaningful engagement should be sought by Governments as well as technology partners/vendors/private partners.

Coordination

For optimum results, streamlining and aligning Government processes, initiatives and information streams at all levels need to be ensured.

This is reiterated in the following (Fig. **9**):

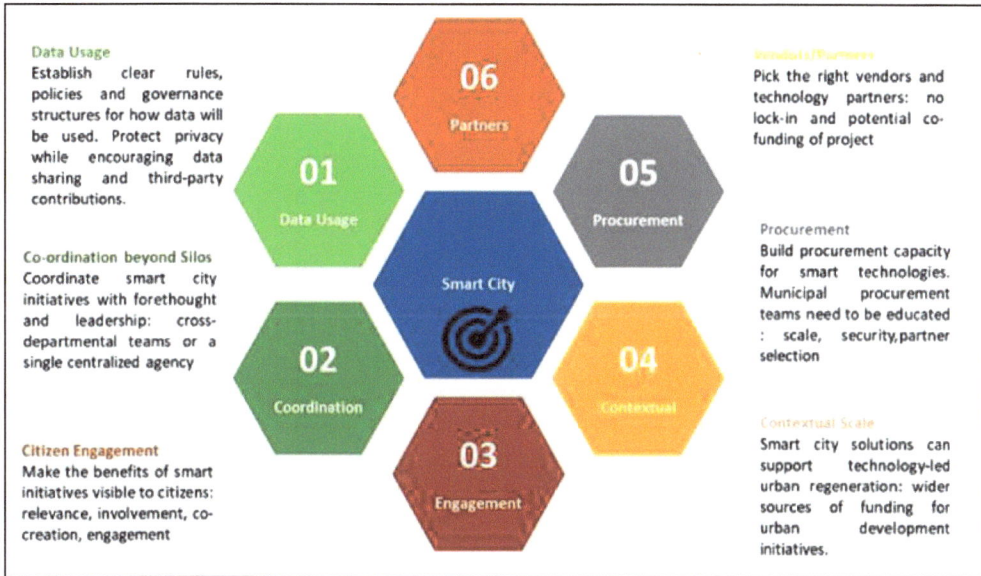

Fig. (9). Smart City Best Practices according to the Smart City Playbook (Source: Machina Research for Nokia http://resources.alcatel-lucent.com/asset/200702) [3].

CHALLENGES IN CREATING AI-ENABLED SMART CITIES IN INDIA

It has been found that while AI applications in smart city create efficiency and automation, they also create regulatory issues pertaining to discrimination in the delivery of services, privacy and ethical and legal issues [7].

The main challenges are listed below:

Maintenance costs and Hardware

High tech instruments like surveillance cameras and sensors require huge resources when put to public service. Recurring costs in the form of maintenance and repairs further add to the expenditure, particularly in a country like India. Creating and maintaining huge Data centres for data storage and processing would be a costly affair though it may lead to employment generation as well.

Privacy Issues

Access to this technology should be strictly controlled as it should not be perceived as an infringement of citizen's right to privacy [13].

Discrimination within City

Care needs to be taken to use AI-enabled services uniformly in all city areas as failure to do so may lead to a feeling of discrimination.

These challenges are evident from the delay in the implementation of Smart Cities Mission in India from 2019-20, to June 2023.

CONCLUSION

While it can hardly be overstated that smart cities would be safer, more energy efficient and well planned, India's cities cannot be compared with the global smart cities of New York or Tokyo as our cities are non-homogeneous and fundamentally different from them. The implementation of smart city is highly context dependent (nations, government *etc.*) [8].The four Indian cities of New Delhi, Mumbai, Hyderabad, and Bengaluru occupy the 89th, 90th, 92nd and 93rd positions in the Smart City Index 2021 conducted for 118 cities and published by the Institute for Management Development (IMD) in collaboration with Singapore University for Technology and Design (SUTD) [9, 12].The top 3 positions went to Singapore, Zurich and Oslo, respectively. Reverse migration witnessed on a large scale in India during lockdown, following Covid-19, ensured that no planning can fit in with the ebb and flow of workers coming and going to and from cities. Workers migrating during sowing and harvesting seasons to their villages, and coming back to cities to take up jobs (mostly daily wage based), during the lean season makes data driven planning difficult as the number of people inhabiting a city at a particular point of time itself keeps changing. Many efforts are being made at the urban local body level, State level and Centre level but most of the times, these are happening in isolation. Integrating these efforts is crucial in creating a powerful synergy within the SCM.

Also, given the typical structure of the Indian economy and the demands on government spending, there is most likely a trade-off between the expenditures made by the government on SCM and those on general public services. Although the SCM has been envisaged as a self-financing system, initial government expenditure amounting to ₹ 2,05,018 crores has already been made. It is therefore important to ensure that corruption is checked and projects are delivered on time.

It has also been pointed out that the entry of private sector in the SCM through PPP (Public Private Participation) and joint ventures cannot ensure inclusiveness and the guidelines do not indicate whether economic efficiency or inclusiveness would take precedence in case of a clash. The process of selection of a city under SCM also needs to be transparent. A case in point is the city of Vijaywada in

Andhra Pradesh which was widely expected to be selected but was not, while non-mission smart cities continue to get support from the Government.

In view of the above, the Government must ensure that urban modernization/AI-enabled SCM is also inclusive.

REFERENCES

[1]   Available at: https://omdia.tech.informa.com/

[2]   Available at: https://www.idc.com/getdoc.jsp?containerId=prAP48288921

[3]   Available at: https://smartcities.gov.in/

[4]   Available at: https://www.worldbank.org/en/topic/urbandevelopment/overview

[5]   Available at: https://www.youtube.com/watch?v=O5xeyoRL95U

[6]   F. Cugurrullo, "Urban Artificial Intelligence", In: *From Automation to Autonomy in the Smart City.* vol. 2. Frontiers: Frontiers in Sustainable Cities, 2020.
[http://dx.doi.org/10.3389/frsc.2020.00038]

[7]   A. Zhou, and Kankanhalli , *AI regulation for smart cities: Challenges and principles, Smart cities and smart governance.* Springer: Cham, 2021.

[8]   F.U. Weisi, and P.E.N.G. Ping, "A discussion on smart city management based on meta-synthesis method", *Manag. Sci. Eng.,* 2014.

[9]   Available at: https://www.imd.org/smart-city-observatory/home/#_smartCity

[10]   Available at: https://www.gartner.com/en/documents/3887767

[11]   V. Jacometti, "Circular Economy and Waste in the Fashion Industry", Available at: www.mdpi.com/journals/laws (2019).

[12]   A. Garg, "Social and Environment Sustainability of Smart Cities", In: *Social and Environment Sustainability of Smart Cities.* vol. 5. IJTIMES, 2019, no. S1, pp. 1-4.

[13]   V. Vashisht, A.K. Pandey, and S.P. Yadav, "Speech Recognition using Machine Learning", In: *In: IEIE Transactions on Smart Processing & Computing.* vol. 10. The Institute of Electronics Engineers of Korea., 2021, no. 3, pp. 233-239.
[http://dx.doi.org/10.5573/IEIESPC.2021.10.3.233]

[14]   Available at: https://mapsofindia.com/

Virtual Reality to Augmented Reality: Need of the Hour in Human Resource Management

Neerja Aswale[1], **Pooja Agarwal**[1] and **Archana Singh**[1,*]

[1] *Faculty of Commerce & Management, Vishwakarma University, Pune, Maharashtra 411048, India*

Abstract: Augmented reality is the need of the hour for Human Resource Management in this era of globalization wherein the world has become flat and businesses have no boundaries. This global business has made varied functions to work in the virtual world full of ease. The functions not only include finance, marketing, logistics, and supply chain but also Human Resource Management has taken a face towards the virtual world for its functions like recruitment, selection, performance management, competency management, or training. The function started with virtual reality and has eventually made a paradigm shift towards augmented reality. The chapter presents the evolution, applications, and challenges of VR and AR with respect to HRM.

Keywords: Virtual Reality, Human Resource Management, Augmented Reality.

INTRODUCTION

Augmented Reality (AR) is an advanced technology that overlays real scenes with computer images. One of the best technical reviews is one that defines the topic, explains many of the issues, and discusses the latest developments. This paper is a good starting point for anyone interested in researching or using augmented reality. AR is part of the collective term mixed reality (MR), which refers to a multifaceted spectrum of topics that include virtual reality (VR), AR, telepresence, and other related technologies [1]. The study demonstrates the potential of VR in HRMD by providing a systematic literature review that focuses on the recruitment, development, and retention of HRM techniques from a scientific perspective. Virtual reality refers to laptop-generated 3D worlds that allow users to enter and interact with simulated settings. Users are prepared to "immerse" themselves to varying degrees into the computer's artificial world, which can be a simulation of a type of reality or a simulation of a posh phenomenon—for example, an AR with virtual seats and a virtual lamp.

* **Corresponding author Archana Singh:** Faculty of Commerce & Management, Vishwakarma University, Pune, Maharashtra 411048, India; E-mail: archana.singh@vupune.ac.in

Adarsh Garg, Valentina Emilia Balas, Rudra Pratap Ojha & Pramod Kumar Srivastava (Eds.)

Telepresence attempts to improve the operator's sensory-motor ability and problem-solving ability in a remote setting [2]. It explains how telepresence works as a human/machine system where the human operator gets enough information about the teleoperator and the task environment [3]. Furthermore, In telepathy tries to create the illusion of being in a different location from virtual reality, where we aim to create a sense of presence in a computer simulation.

Virtual reality and perspective can both be used to describe augmented reality. While the environment in virtual reality is fully made up and appears authentic in the scenario, in augmented reality, the viewer sees the actual world enhanced with virtual things. Three factors must be taken into account while creating an AR system:

1. The combination of natural and virtual worlds;
2. Interaction- Real Time;
3. 3D recording.

Augmented scenes can be displayed *via* wearable devices such as head-mounted displays (HMDs), but alternative technologies are also available. In addition to the three previously mentioned qualities, portability can be included. Due to device restrictions, most virtual environment solutions do not allow users to move around much. Some AR applications, on the other hand, need the user to walk through a larger environment. As a result, portability becomes a critical concern. 3D registration becomes considerably more complicated for such applications. Mobile computer apps frequently provide text/graphic information that the monocular HMD does not capture.

Computing platforms and wearable displays utilised in Augmented Reality must now frequently be adapted for broader applications. The field of augmented reality has been established for over a decade, but recent years have seen significant growth and progress. The field has advanced tremendously since then. Several conferences specialising in this field, such as the International Augmented Reality Workshops and Symposium, the International Conference on Mixed Reality, and the Augmented Reality Environmental Conference, have been launched [4].

AUGMENTED REALITY IN HUMAN RESOURCES MANAGEMENT

It combines technology, computer vision, and interaction to make intelligent solutions for employees. Using computer vision to collect information from various sources, like videos, images, text, and voice, can identify template trends and inform employees of the tasks they need to perform. Computer vision should include the artificial or natural picture, during which employee movements are captured by camera sensors, face recognition technology, or fingerprint readers. The info is then analysed to give an overview of trends [5]. The field of augmented reality in people management can be defined as the use of technology, computer vision, and interaction to produce intelligent solutions for employees. Using computer vision to gather information from many sources, such as video, photos, text, sound, and speech, it is possible to discover trends in the workforce and alert employees of essential duties. Computer vision must include a machine or realistic picture recording, employee movements by camera sensors, facial recognition technology, or fingerprint scanning. The data is then analysed to provide insights into trends [6].

Integrating virtual reality and augmented reality is one-way employers can use these technologies to increase productivity and efficiency and reduce labor cost. Employees are already aware of the benefits of using e-learning software and are adopting new technologies to improve their productivity. Employees are trained on specific tasks using augmented Reality in HR departments located in virtual offices involved in all business activities, from conference calls to group training sessions. Organizations looking to reduce costs associated with hiring, training, and training employees can also benefit from using these technologies. Digital Signage, which involves taking pictures with a camera or another device and displaying them on a small screen, is one instance of an AR utilized in HR. Employees get access to well-known programs like Google Maps, which allow them to share addresses, find addresses, and connect with friends on the renowned Google+ social network. Using popular augmented reality, apps may bring a second layer of reality to your company, allowing your staff to view a virtual representation of their actions in real-time. Although augmented reality (AR) has been talked about for a while, it has only recently gained traction. Several businesses already use this technology in their daily operations. Even Google now offers it's mapping services with this technique. AR has numerous possible applications. Now let's talk about a few of them [7].

EVOLUTION OF AR AND VR

In the 1920s, the first attempts were made to develop technology to demonstrate augmented and virtual reality. AR and VR technologies emerged in the mid-to late 20th century. Products with this technology are now available from a variety of large and small retailers. Virtual reality headsets are often designed to look like masks, goggles, or other types of facial clothing. In 1960, cinematographer and VR pioneer Morton Heilig invented the telesphere for his mask, which became his first-ever head-mounted display. (HMD). Telesphere Mask uses stereo techno-logy, 3D imaging, widescreen vision, and stereo sound to simulate virtual reality for the user. The SAYRE GLOVE is the first connected glove with AR/VR technology, developed in 1977 by scientists at the University of Illinois Electronic Visualization Laboratory. Power gloves and data gloves were introduced in 1982. Both gloves are from Thomas G. Zimmerman and Jaron Lanier and used optical flex sensors and ultrasonic and magnetic hand position tracking techniques. AR/VR technology soon found its way into arcade machines and video games. Even more interestingly, NASA and Nintendo acquired the technology needed to create simulations of video game consoles and body suits. Today's market is flooded with AR and VR devices.

Google Cardboard ($15) offers an affordable VR headset simulation. Although the future of virtual reality is unknown, augmented reality and virtual reality are changing the way we interact with the world. Technology permeates education, engages students in virtual field trips, language immersion, and game-based learning, provides hands-on real-time project collaboration and virtual assistance, and extends to field service operations. These technologies add to professional training and make medicine more accessible to health systems. In 1838, Charles Wheatstone c used the image of each eye to create a three-dimensional perspective display, leading to virtual reality, and augmented reality. Since then, technology has evolved and become more integrated into our daily lives.

XR: Augmented reality, or XR for short, refers to all immersive technologies such as augmented reality, virtual reality, and mixed reality. All immersive technologies allow you to interact with a mix of natural and virtual worlds or virtual and immersive reality.

AR: Augmented reality, often called AR, allows virtual products and information to be placed in a virtual environment. AR glasses or any digital display such as smartphones and tablets can be used by the user. By using these objects, a person is not separated from the real world and can still communicate and observe what is happening around them. A well-known example of AR is the game Pokémon GO, which places digital beings into the real world, or Snapchat filters, which place digital objects on the head, such as hats or glasses.

VR: VR or virtual reality completely immerses the user in a virtual world. To get a full 360-degree perspective of the virtual experience, consumers must use a VR headset for these sessions. Advanced technology in the gaming and entertainment industry is gradually making its way into other sectors such as medicine, construction, and more [8].

ROLE OF VR & AR IN HUMAN RESOURCES

When the COVID-19 pandemic struck, certain industries were impacted much harder than others, particularly the service and entertainment industries, which experienced a near-complete shutdown and struggled to bring employees back in the months that followed. Laura Lee, CHRO of MGM Resorts, declared during a recent HRE webinar, "The Great Recession is real." "We had no idea we'd be closing down every casino on the Las Vegas strip." MGM has made significant investments in its staff, enabling greater employee engagement and assisting in the development of their abilities and confidence to give the most outstanding customer experience possible through virtual reality [9].

Derek Belch, the founder and CEO of Strive, compares virtual reality to pilot training. Pilots can practise flying in virtual reality until they can demonstrate that they can do so in real life. "VR allows all learners—from janitors to CEOs—to learn by doing," says Belch, who also attended the webinar. Other HR executives are exploring adopting similar VR training tools in a small department or division to begin with. "Find and scale that one fabric that pervades your business," she suggests. "We worked in guest services." [10]

"Collaborate with your team to determine the ideal conditions. If necessary, grab a COO or the head of hospitality and attach some headsets to them. They'll figure it out eventually. That is critical." And don't be scared. Understanding your company's strategy and workforce is crucial, especially for HR executives. They should always try to expand learning opportunities so that employees can perform to their full potential [11].

In human resources, augmented reality refers to incorporating virtual features such as picture displays, synthetic data, and speech recognition into the hardware of existing enterprise software to improve its effectiveness and usability. Business owners and managers may make educated decisions about staff performance by having real-time access to critical data. In this manner, you may avoid wasting training time and save money. By integrating smartphones, tablets, and other handheld devices, augmented reality will allow your staff to access and display the information they require in real-time [12].

APPLICATIONS OF AR AND VR IN DIFFERENT HRM FUNCTIONS

Profound changes are happening in the field of Human Resource Management. The different practices and functions are also getting developed through AR & VR. Some of the applications of this technology to HRM functions can be explained as follows:

1. AR & VR in Recruitment and Selection- Talent acquisition is one of the essential and expensive functions of an HR department. The use of AR & VR reduces its various challenges. Companies like general mills used a 3D video to get candidates' attention at a busy job fair in 2015. Similarly, Toyota began collaborating with InstaVR in 2017 to provide virtual office tours that provide virtual access to all office amenities and locations. Holger Muller, on the other hand, believes that VR is not widely used in recruitment functions. Nonetheless, making videos about it has become a habit in recent years. The company may show itself more effectively and efficiently *via* AR and VR [13].

2. AR & VR in the scrutiny of candidates- It is impossible, but the reality is that AR and VR allow employers to screen new employees by replicating and analysing job-related skills. Assessing talent with AR and VR is more competent and effective than in the past. Traditional hiring or shortlisting of candidates is based on their credentials. Still, in the time of AR & VR, it is based on the candidate's skills and talent, which these technologies can measure without offering a job to the candidate [14].

3. AR Based Onboarding- This is also one of the essential functions of human resource management. Effective onboarding increases the happiness and satisfaction of the employee. It also reduces the duration of task completion. In Eckert's words, with AR & VR in onboarding, it is becoming easier compared to earlier. During the pandemic, when people felt isolated and disconnected, they felt good about AR-based boarding.

4. AR & VR in Employee Training- AR & VR is also a revolution in employees' training areas. It provides a platform to learn new skills in a short period with the use of AR & VR. For example, Walmart established a space where employees can learn new goods pickup skills with the help of AR & VR, which narrowed down the time from 8 hrs to 15 minutes. VR and AR also assist in learning soft skills like leadership, interpersonal, diversity, equity, inclusion, *etc* [15].

COMPANIES USING AR & VR

The application of AR and VR in HRM functions has been imbided in varied HR functions in many companies. In the same way, different business organizations are using this technology differently as per their requirements and offering better products and services than before. Organizations having the application of AR and VR in their business are summarized in the following Table **1**.

IMPLICATIONS AND DIFFICULTIES OF VR AND AR IN HUMAN RESOURCES

Many technological advancements are taking place in the present era of Human Resource Management (HRM). The people management routine is becoming increasingly demanding daily. As a result, HR managers are constantly on the lookout for technologies that can improve the working compatibility of HR operations. For this, recent technology known as Virtual Reality can be favoured by HR management for recruiting personnel, database administration, and boosting staff productivity and communication, especially if it is impactful and simple to use. Virtual reality (VR) fulfills both of these requirements. VR is used by both large corporations and agile start-ups for everything from hiring to training to employee communication. The virtual platform they picked allows their staff to learn from one another and to benefit from language acquisition. Hence, we studied various studies and research on virtual concept and pointed out the objectives, challenges, and future of this Virtual HR concept [20, 21].

Table 1. Application of AR & VR by Different Companies.

S. No.	Sector	Company Name	Application of AR & VR by Different Companies
1.	Retail	Nike	This company uses AR & VR at their shopping outlets. With this application, customers can scan their products like shoes or clothes to gather entire information about them.
		IKEA	This company has developed the place app which facilitates customers to apply this with their smartphone camera to place furniture items in their homes to visualize the look of the furniture in their home settings.
		Warby Parker	This company uses AR to provide a facility to customers to try glasses by sitting at their homes and choosing the best frame that suited them without visiting the outlet and ordering accordingly.
		L'Oreal	L'Oreal is one of the known companies for beauty products which are using AR and VR in giving the privilege to their customers to see their looks after applying these beauty products virtually not in reality.
		BMW	This company is using the concept of AR and VR widely. With the use of this technology, the customer can visit BMW's showroom and can make customized changes (color or style) in the model by using their tablets or phones [16]. In another way, the company is using this technology by making the customer wear virtual reality goggles to make them feel about the driving experience. So, they can analyze their options and can select the best option.

S. No.	Sector	Company Name	Application of AR & VR by Different Companies
2.	Education	Boulevard	The approach of this company is unique. With the help of AR & VR, this company makes it possible for art lovers to visit different places, art galleries, or museums without moving from one place to another. This company has collaborated with some prominent museums in UK and USA and facilitated people to experience all with AR & VR.
		Nearpod	This company utilises AR and VR in novel ways to create more interesting and spectacular classrooms. They identify the underlying value of traditional lesson plans and create a win-win situation for both parties. Educators can employ integrated 360-degree visuals and videos to make lessons more engaging, and students can engage in interactive learning even when they are at home [17].
3.	Health Care	AccuVein	This company is making life easier for patients as well as nurses. A marketing survey quoted that 40% of intravenous injections miss the vein, which is highly problematic for kids and elderly people. With the help of AR & VR, medical professionals use a handheld scanner that projects blood clots over the skin. It highlights the veins in the patients' bodies.
		Curiscope	This company comes up with an innovative idea in the form of a Virtual-tee T-shirt. This T-shirt is not a normal one, but it is an application based on AR & VR through which a person can see the anatomy of the human body through realistic holograms [18].
4.	Others	GOOGLE	With the use of AR & VR, Google Maps make it easy to go in interior complicated areas also with the navigation by this.
		Airbus	AR and VR are also used in the manufacturing and engineering field.. The company Airbus is using this technology to increase the efficiency of its aircraft by taking a virtual experience to understand the comfort label for passengers and examine the requirement of safety standards [19].

CONCLUSION

AR and VR are a boon for today's industry, but on the other side, sometimes they reveal some drawbacks. Firstly, this technology uses long-term VR headsets, which is ok for youngsters because they spend a lot of time in the virtual environment. On the other side, aged or experience folks experience motion sickness, dizziness, and other issues. Secondly, this is an expensive way to learn, which poor people cannot afford. It also creates inequality in education. Thirdly, it is becoming a reason for addiction in the young generation. Youngsters were abducted for this, which resulted in inhumanity. Technology Advancement has brought us a long way, but the human touch is immensely needed and specially for HR. The need of the hour is still technology-based HRM as slowly and steadily it is gripping all businesses and all functions and yes HR cannot be left behind and is benefitting in all terms through VR & HR.

REFERENCES

[1]   K.M. Baird, and W. Barfield, "Evaluating the effectiveness of augmented reality displays for a manual assembly task", *Virtual Real.,* vol. 4, no. 4, pp. 250-259, 1999.
[http://dx.doi.org/10.1007/BF01421808]

[2]   P. Bergeron, "Finding the Value in Virtual Reality for HR SHRM Executive Network", Available at: https://www.shrm.org/executive/resources/articles/pages/value-of-virtual-reality-bergeron.aspx(2022).

[3]   Brown, "Augmented reality – the next big thing for HR? HRD Connect", Available at: https://www.hrdconnect.com/2019/07/26/augmented-reality-the-next-big-thing-for-hr-2/(2019).

[4]   J. Carmigniani, and B. Furht, "Augmented Reality An Overview", In: *Handbook of Augmented Reality.* Springer, 2011, pp. 3-46.
[http://dx.doi.org/10.1007/978-1-4614-0064-6_1]

[5]   D. Cliburn, S. Rilea, J. Charette, R. Bennett, D. Fedor-Thurman, T. Heino, and D. Parsons, "Evaluating presence in low-cost Virtual Reality display systems for undergraduate education", *J. Comput. Sci. Coll.,* vol. 25, no. 2, pp. 31-38, 2009.

[6]   P. Ferreira, V. Meirinhos, A.C. Rodrigues, and A. Marques, "Virtual and Augmented Reality in human resource management and development: A systematic literature review", *IBIMA Business Review,* vol. 2021, pp. 1-18, 2021.
[http://dx.doi.org/10.5171/2021.926642]

[7]   T. Haak, "Virtual reality and augmented Reality: Why HR and People leaders should embrace VR and AR Sage-People and Leadership", Available at: https://www.sage.com/en-gb/blog/virtual-reality-vr ar-tips-hr/(2020).

[8]   J Krithika, P Venkatraman, and EA Sindhujaa, "Virtual hr era in human resource management", In: *EPRA International Journal of multidisciplinary Research.* vol. 5. IJMR, 2019, no. 10, pp. 2455-3662.

[9]   S.P. Yadav, and S. Yadav, "Fusion of Medical Images in Wavelet Domain: A Discrete Mathematical Model", In: *In Ingeniería Solidaria.* vol. 14. Universidad Cooperativa de Colombia: UCC., 2018, no. 25, pp. 1-11.
[http://dx.doi.org/10.16925/.v14i0.2236]

[10]   G. Lawton, "VR in HR: How human resources can use VR and AR technology. Tech Target", Available at: https://www.techtarget.com/searchhrsoftware/tip/VR-in-HR-How-human-resources can-use-VR-and-AR-technology(2021).

[11]   R. Mehtab, "R and VR: New Landscapes in Human Resource Industry Linkedin", Available at: https://www.linkedin.com/pulse/ar-vr-new-landscapes-human-resource-industry-rida-mehtab/(2021).

[12]   M.A. Muhanna, "Virtual reality and the CAVE: Taxonomy, interaction challenges and research directions", *Journal of King Saud University - Computer and Information Sciences,* vol. 27, no. 3, pp. 344-361, 2015.
[http://dx.doi.org/10.1016/j.jksuci.2014.03.023]

[13]   A. Nikolaidis, "What Is Significant in Modern Augmented Reality: A Systematic Analysis of Existing Reviews", *J. Imaging,* vol. 8, no. 5, p. 145, 2022.
[http://dx.doi.org/10.3390/jimaging8050145] [PMID: 35621909]

[14]   V. Vashisht, A.K. Pandey, and S.P. Yadav, "Speech Recognition using Machine Learning", In: *IEIE Transactions on Smart Processing & Computing.* vol. 10. The Institute of Electronics Engineers of Korea., 2021, no. 3, pp. 233-239.
[http://dx.doi.org/10.5573/IEIESPC.2021.10.3.233]

[15]   S. Ronan, "How augmented reality is infiltrating the world of HR People Management", Available at: https://www.peoplemanagement.co.uk/article/1742209/augmented-reality-infiltrating-world-hr(2020).

[16]   L. Muñoz-Saavedra, L. Miró-Amarante, and M. Domínguez-Morales, "Augmented and Virtual Reality Evolution and Future Tendency", *Appl. Sci. (Basel),* vol. 10, no. 1, p. 322, 2020.

[http://dx.doi.org/10.3390/app10010322]

[17] G Santi, A Ceruti, A Liverani, and F Osti, "Augmented reality in industry 4.0 and future innovation programs", *Technologies.,* vol. 9, no. 2, p. 33, 2021.
[http://dx.doi.org/10.3390/technologies9020033]

[18] R. Sharma, and S. Sabat, "AR/VR -the way Forward in HR Management Business and Technology Insights", Available at: https://www.tcs.com/blogs/ar-vr-the-way-forward-in-hr-management(2020).

[19] D.W.F. Van Krevelen, and R. Poelman, "A survey of augmented reality technologies, applications and limitations", *Int. J. Virtual Real.,* vol. 9, no. 2, pp. 1-20, 2010.
[http://dx.doi.org/10.20870/IJVR.2010.9.2.2767]

[20] E. Vasilenko, "Virtual reality in HR management as a condition of innovative changes in a company", *International Conference on Digital Transformation in Logistics and Infrastructure (ICDTLI 2019) Atlantis Highlights in Computer Sciences.,* vol. 1, 2019.
[http://dx.doi.org/10.2991/icdtli-19.2019.85]

[21] "Visualização de De e-GDRH a v-GDRH_ revisão sistemática da literatura da Realidade Virtual e Aumentada na Gestão e Desenvolvimento de Recursos Humanos".

CHAPTER 11

AI-enabled Innovations and Green Economy in Fashion Industry

Adarsh Garg[1,*] and **Amrita Jain**[1]

[1] *Data Analytics, G. L. Bajaj Institute of management and Research, Greater Noida, India*

Abstract: Digitization has a substantial impact on almost all facets of fashion, starting from the designing of a fashion item to its production and its usage by consumers. Fashion has always been evolving with emerging technologies. With the beginning of Artificial Intelligent (AI) enabled technologies, the fashion industry has become as dynamic as technology, emerging as a forward-looking trend giant. The impact of AI on the augmentation of fashion trends is unquestionable and the industry has witnessed its fast move from 4.0 to 5.0 with the use of advanced technology. Although fashion is changing at a very fast pace with AI, fashion professionals have raised the socio-economic impact of AI on the fashion industry, including the Green Economy (GE) issues, thus, making the exploration of the phenomenon essential. This chapter explores how AI-enabled technology in the fashion industry and fashion environment, is influencing the GE status of the fashion industry, especially in the post-COVID-19 era of innovative e-commerce fashion.

Keywords: Fashion design, Digitalization in Fashion, Artificial Intelligence (AI), Green Economy (GE).

GREEN ECONOMY

The United Nations Environment Programme (UNEP) defines "an improved human well-being and social equity, while significantly reducing environmental risks and ecological scarcities" or "a low carbon, resource efficient and socially inclusive economy", which is known as a green economy.

In a green economy, investments that reduce greenhouse gas emissions and pollution, promote energy and resource efficiency, and stop the deterioration of biodiversity and ecosystem services should be the driving force behind the job and income creation. A "green economy" is one that aims to reduce environmental dangers and ecological scarcities and pursues sustainable development without damaging the environment [27].

---

* **Corresponding author Adarsh Garg:** Data Analytics, G. L. Bajaj Institute of management and Research, Greater Noida, India; Tel: 91-9818627629; E-mail: adarsh.garg@glbimr.org

The green economy is the origin of growth that is built on an innovative network of economic activities that support social well-being, and wholesome natural ecosystems, and result in successful yet moral company expansion. Life, earth, and income are the three pillars of sustainable development, and the green economy encourages change and progress in those directions. The production and use of renewable energy, energy efficiency, waste minimization and management, preservation and sustainable use of already existing natural resources, and the creation of green jobs are the main components of the green economy.

India stepped down from 21st to 42nd place in the Green Future Index 2021 to 2022. Whereas, Iceland, Denmark, and the Netherlands are the countries most prepared for a low-carbon future, according to a new report. Other countries making up the top 10 of the Green Future Index 2022, are the United Kingdom, Norway, Finland, France, Germany, Sweden and South Korea [28].

NEED FOR GREEN ECONOMY IN THE FASHION INDUSTRY

The worldwide fashion industry produces wastes, and pollution, and uses a lot of energy. The fashion industry hasn't yet taken its environmental duties seriously enough, despite some modest advances. In order to meet the customer's desire for a revolutionary change, the fashion industry's players will need to move away from clichés and advertising noise toward genuine action and regulatory compliance.

The fashion industry flourished in the early twenty-first century. Apparel manufacturing doubled between 2000 and 2014, while the average customer bought 60% more clothing per year as a result of declining costs, streamlining processes, and rising consumer expenditure [1]. Total industry revenues are anticipated to increase annually starting in 2021. In reality, according to the market's 2023 prediction, revenues will rise 5.48 percent to slightly over $1.8 trillion. Given that the average annual growth rate for the global apparel market between 2013 and 2026 is 2.24 percent, this is a significant increase. For some garment companies, fast fashion has been a particularly popular market and a source of remarkable growth.

Sales of apparel increased eight times more quickly in five major developing nations than they did in Canada, Germany, the United Kingdom, and the United States: Brazil, China, India, Mexico, and Russia as depicted in (Fig. **1**).

Change in consumer prices, 1995–2014, %

■ All goods   ■ Clothing

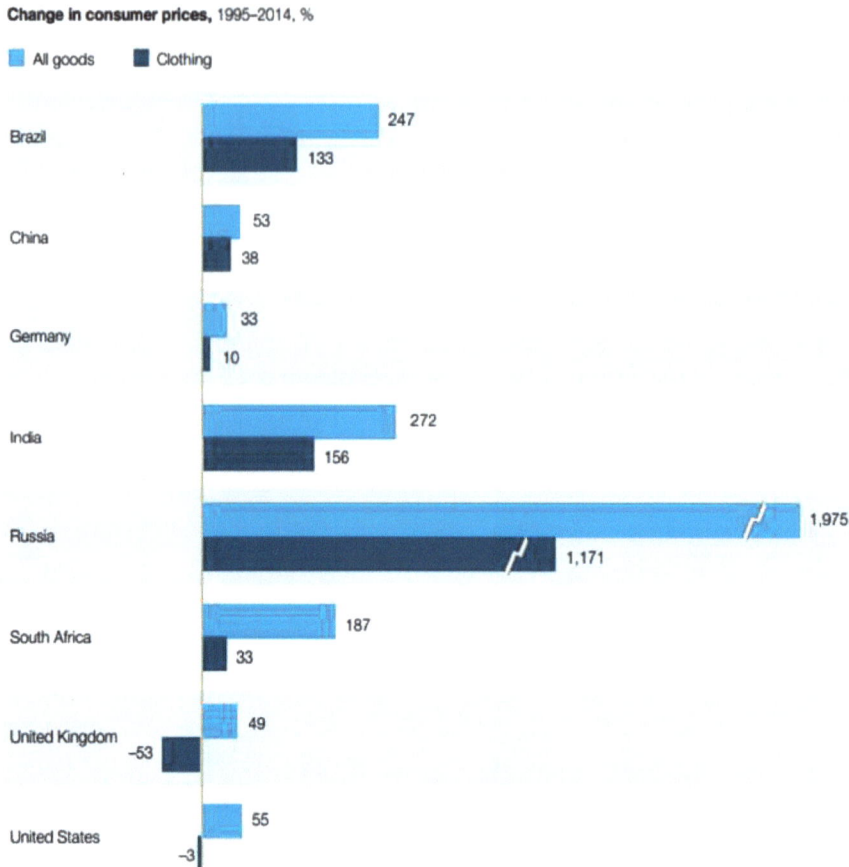

Brazil — 247 / 133
China — 53 / 38
Germany — 33 / 10
India — 272 / 156
Russia — 1,975 / 1,171
South Africa — 187 / 33
United Kingdom — 49 / -53
United States — 55 / -3

Fig. (1). Change in Clothing Expenditure. Source: McKinsey & Company report, 2016.

Cotton is often farmed using a lot of water, pesticides, and fertiliser, making up around 30% of all textile fibre use. We calculate that the production of 1 kilogramme of fabric results in an average of 23 kilogrammes of greenhouse gases since nations with significant fabric and clothing manufacturing industries mostly rely on fossil fuels for energy production (Fig. **2**).

Increases in environmental impact if 80% of emerging markets achieve Western per capita consumption levels[1]

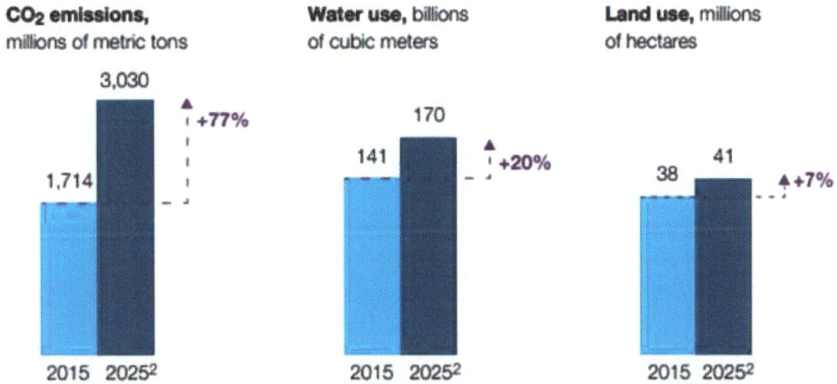

Fig. (2). Environmental Impact of Fashion Industry. Source: McKinsey & Company report [31].

After customers leave the store with brand-new items, the environmental effect of clothing keeps growing. Our estimations show that washing and drying 1 kilogramme of clothes over its full life cycle produce 11 kilos of greenhouse emissions, a number that businesses may lower by changing textiles and garment designs. Consumer decisions made after the fact, such as whether to wash clothes in cold, warm, or hot water, have a significant impact.

Current technologies cannot consistently convert discarded clothing into fibres that may be used to create new products in regard to disposing of clothing. Recycling techniques like chemical digestion and shredding do not work effectively. Additionally, there aren't enough markets to take in the amount of material that would result from recycling clothing. As a result, within years of production, almost three-fifths of all apparel manufactured are disposed in dumpsites. By collecting roughly three-quarters of all used clothing, reusing half, and recycling the remaining quarter, Germany beats the majority of other nations. Collection rates are far lower in other countries: 15% in the US, 12% in Japan, and 10% in China.

TECHNOLOGY IN FASHION INDUSTRY

With Industry 4.0, the digital revolution is allowing the fashion industry to increase its product line very rapidly with consumers and manufacturers both more focussed on data as information. A new fashion mass market has evolved with more creativity, innovations, and technology-led applications, making a paradigm shift from measure-centric approach of fashion to style-centric ready-to -to-wear approach. These technological innovations are not restricted to style or speed or customization or financial feasibility or logistics, but a lot has been explored in unexplored fabric areas also.

A large number of innovations and patents are developed. These innovations significantly improved the manufacturing and consumption of clothes. The technological advances gave birth to the modern fashion industry. Industry 4.0 conceptualized rapid changes in technology, industry, and patterns of society. This change is followed by industry 5.0 with humans working alongside smart machines, robots [2, 3]. Primarily, AI is the major tool offering solutions for undertaking difficulties of sustainability in the fashion and retail industry. However, the foundation were laid years ago with the introduction of technologies in the manufacturing process, transforming a small manufacturing system to a large industry.

Fashion industry, which is changing and progressing continuously, faces new challenges almost every day. Today, it is one of the foremost economies in the world. It has an estimated worth of up to 3,000 billion US dollars [4, 5]. As mentioned earlier, cloth manufacturing is one of the oldest social happenings. Since then, it has evolved continuously through the centuries with the adaption of technology and society advances. The fashion industry is rapidly adopting 5.0 cloth production technology, along with all latest digital accomplishments. Today, the world is more interested to have a holistic advancement in terms of economic, social, and cultural sustainability. The major goal is to enhance activities that eliminate the polluting impact of industry and fashion industry among such industries. With the ever increasing and diversified demand from consumers of fashion products, it has somehow created a lot of waste. Last decade has witnessed significant digital innovations not only in sales and retail but also in the digitalization of the whole supply chain and design in the fashion industry to revitalize it with the help of AI. However, when technology is functional in an uninhibited way, it somehow accelerates the already exploited economic progression. Thus, disrupting the sensitive commercial arrangement of the fashion industry. A number of initiatives have been taken to reduce the environmental impact of the fashion industry, to optimistically move towards a green economy.

AI IN FASHION INDUSTRY

The global expenditure on AI in Fashion will grow from USD229 million in 2019 to USD 1260 million by 2024, according to CAGR of 40.8 percent Report [6]. A lot has been attributed to the dynamic and global connect between buyers and sellers, with the help of technology [7] with exceptionally high activity during pandemic 2019. The pandemic has enhanced the online activities in the fashion industry multifold due to obvious confinement indoor [8]. Most of the business processes became data-driven in the fashion industry. From the trend analysis, logistics sales forecasting, online shopping to fashion recommendation have got a major innovative boost with AI- enabled process [9].

The consumer buying behaviour has observed a totally tailored and customized style [10]. As of now, a variation in innovative applications can be seen. For example, ViSENZE [11] gives fashion image processing applications. This helps in searching an image and predicting the attributes. Zalando team is looking for fashion goods recommendations [12]. Some more innovative applications include Intelistyle [13] with a chatbot-based AI inventor, and a wardrobe-based AI inventor Fitzme [14].

AI has changed the working of each industry, as well as fashion. With almost escaping of the physical mode of shopping, e-retail has become the choice [11]. At this time, AI-enabled solutions would also support realizing sustainability goals in the fashion industry, to promote a green economy, as it can be specified that people have not ever consumed as many apparel as today. Most of the AI technologies used in the fashion industry, so far, can be listed as follows:

FASHION DESIGNING

Fashion is an advent of the human imaginary vis-à-vis the human body. This includes the body shape for clothing and creativity which is addressed by inventors and artists. It involves a lot of intelligence, brainstorming, and decision-making. Fashion experts are doubtful if such a complex process, named fashion design, can be automatised as shown in (Fig. **3**).

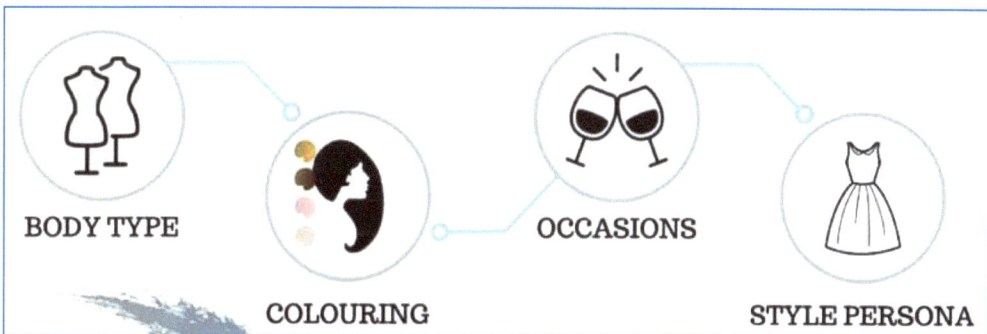

Fig. (3). Fashion Design Process.

The fashion industry has already experienced simple but competitive AI-enabled development with the launch of Lab126 in 2004 by Amazon, which has the ability to trace images and generate a new style. Social media postings are speedily tracked by e-retailers for revolutionary fashion trends. These retailers take the help of smart and emerging technologies to ensure the latest trend design with precision, increased speed of manufacturing, the flexibility of logistics, lower cost, and highly customized products with virtual feel and acceptance by diversified customers, *e.g.* Lenskart, and Nike shoes.

It has further competence in the sustainable economy by reducing the wardrobe size and increasing the satisfaction level of customers with the latest trendy clothing collections.

FORECASTING IN FASHION

Some AI-enabled hybrid tools have been established as very effective in the prediction of fashion sales [15, 16]. In today's very fast-paced market, it is quite challenging to get to the increasing demand of customers. Earlier, the companies were dependent on reports and trend books for designing various apparel for their potential customers. Now it is more real-time data-oriented. . Forecasting of fashion becomes more preferred in terms of the social, cultural, lifestyle, and economic status of customers and this is a vital subject for any fashion brand and to retain the market.

However, predicting successful and sustainable fashion goods is tough, keeping in view the short life cycle of goods and the higher variability of trends. In such situations, the technology is quite helpful. Artificial Intelligence with the ability to foresee trends has/is helping to develop techniques to predict trends by using quality data in a real-time manner. The data comes from different sources like social media, virtual 3D, geographic, point of sale, and logistics data. Emerging technologies are helping retailers by exploring all these types of data for knowledge discovery to manage catalogue, and shelves using AI-enabled tools to measure demand. Thus, the precise measurement of demand helps to promote a green economy [29, 31].

STYLE APPLICATIONS

As mentioned above, AI is essentially renovating the fashion industry from the way of manufacturing fashion goods to the way of selling them. AI technologies are transforming the fashion industry in a way to make the cloth spending experience of the customers ever contented. For example, a virtual DressingRoom app by GAP, which was introduced in 2008, supported customers to virtually try apparels without the need for a physical trial room as shown in (Fig. **4**) [17].

Similarly, in (Fig. **5**), the Nike Fit app is scanning the customer's foot to analyse the best-fit design shoes for the target customers [18]. These applications reduce waste and enhance the economy [30].

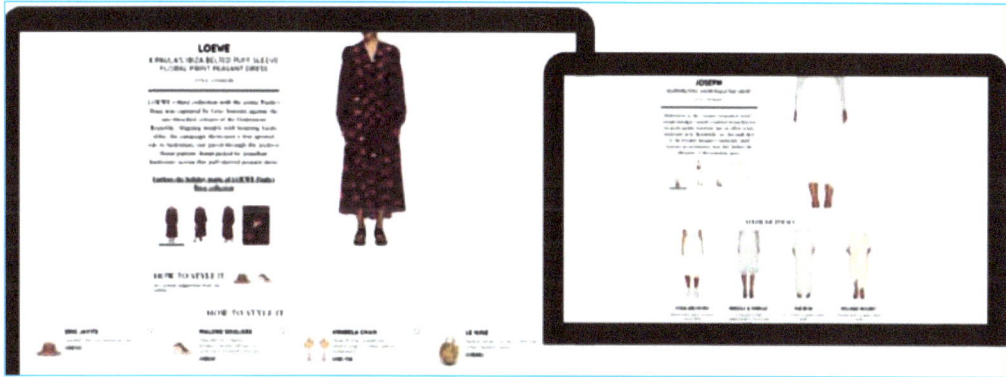

Fig. (4). Virtual trial DressingRoom by GAP.

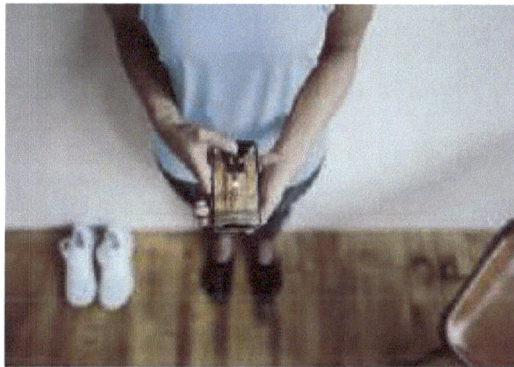

Fig. (5). Virtual Nike Fit Scanning App.

VIRTUAL MERCHANDISING

Further moving, it is quite unavoidable for the customer to buy n number of garments when looking for a style-fit, color-fit, body-fit, or trend-fit. This makes the bulging wardrobe which indirectly impacts the sustainability of the economy. With the advent of e-commerce, people feel more comfortable buying online. All these online businesses, give enough leverage to the customer to happily select the garments by providing the images and information. However, the depiction of the suitability of the garment as per the body dimensions is not provided most of the time. This leads to wastage. Artificial and Virtual reality(AVR) technologies are addressing this problem and reducing the customer experience gap of buying online.

EFFECTS OF THE FASHION INDUSTRY ON ENVIRONMENTAL DAMAGE

Geneva Environment Network [19] elaborates on the damaging effects of the fashion industry on the environment as:

a. The fashion industry emits approximately 8–10% of aggregate greenhouse emission gases, which is highly alarming due to its effect more than the combined emission of international and maritime ships [20]. This is likely to reach 26% by the end of 2050 [21].
b. Every year, approx. 93 billion m^3 of water is used for the production of garments in the fashion industry [22] and every second dumping is equivalent to one truck [23].
c. The fashion industry contributes 20% of industrial wastewater pollution [24]. 60% of the garment material is plastic [25].
d. Washing clothes releases almost 500,000 tons of microfibres annually, which is the same as disposing of 50 billion crumpled plastic bottles in water [26].

Looking at the negative/challenging features of the fashion industry and the way AI technologies are filling the gap, it is essential; to look for a sustainable design in a more holistic manner.

A SUSTAINABLE DESIGN FOR THE FAST-FASHION VALUE CHAIN

Action from the entire industry will probably be needed to reduce the negative effects of the fast fashion industry on sustainability. Some clothing firms have joined together to address environmental and social issues, which has sped up the change and reduced the risks associated with tackling these issues on one's own. An organisation called Zero Discharge of Hazardous Chemicals, for instance, brings together 22 apparel companies to advance and broaden the use of nontoxic, sustainable chemistry in the textile and footwear supply chains. More than 50 brands and retailers, as well as close to 700 suppliers, are involved in the Better Cotton Initiative, which set standards for cotton production's economic, social, and environmental responsibilities [27 - 31].

A few apparel companies have started to take on sustainability issues independently. I:CO has worked with H&M and Levi's to collect garments and footwear for recycling and reuse. I:CO supplies the collection bins, sorts the materials so that anything that may be worn can be sold, and recycles the remainder. In addition to collecting used clothes in its stores and *via* mail, Patagonia also provides repair services to help its customers prolong the life of their apparel.

Additional actions that businesses can take to reduce some of the social and environmental problems that are frequently associated with the fast-fashion model include the following:

- Create guidelines and procedures for designing apparel that is simple to recycle or reuse. An index for evaluating the impact of clothing and footwear goods across their whole lifecycle has been developed by the Sustainable Apparel Coalition.
- Make investments in the creation of newer fibres that will lessen the negative environmental effects of manufacturing and clothing. In order to assist research on increasing the efficiency and sustainability of the textile manufacturing industry, the Walmart Foundation gave grants totalling around $3 million to five US institutions in 2016.
- Encourage customers to take low-impact care of their clothing. It consumes a lot of energy to wash and dry clothes in hot or warm water at high temperatures or for longer than necessary. Clothing manufacturers and retailers should encourage customers to use clothing care techniques that are less harmful to the environment and prolong the life of their clothes.
- Encourage the development of technology for mechanical and chemical recycling. For instance, mechanical recycling results in shorter, lower-quality, and less usable fibres for the apparel industry than virgin fibre. This could be improved by chemical recycling as the technology develops.
- Establish systems to increase supply chain transparency and set tougher labour and environmental requirements for vendors. For instance, the packaging manufacturer Avery Dennison and the software provider EVRYTHNG have joined forces to tag garments so that consumers can track the production of specific goods throughout the supply chain.
- Offer suppliers support and provide resources to help them comply with new labour and environmental standards, and hold them responsible for performance lapses. For instance, Walmart has publicly stated that by 2017, at least 70% of the products it purchases from suppliers directly will come from facilities with energy management strategies. The business provides its suppliers with software tools to assist them in identifying opportunities for more effective use of resources, including energy.

CONCLUSION

With the ever increasing indulging of fashion apparel in today's digital era, and the increased application of Artificial Intelligence in fashion, new questions are being raised pertaining to the quality of design and its social aspects. Onset of e-commerce has given the biggest challenge to sustainability. The chapter tries to summarise the opportunities that AI is creating in the fashion industry and explored further possibility of addressing the sustainability challenges.

REFERENCES

[1]   "McKinsey Style that's sustainable: A new fast-fashion formula", 2016.

[2]   P. Cochrane, "Excellence in Sustainability Gives Turkey Clothing Sector Cutting Edge in Global Markets", Available at: https://www.just-style.com/thought_leaders/sustainability-turkey-cloth-ng-sector-global-markets(2021).

[3]   C. Bai, P. Dallasega, G. Orzes, and J. Sarkis, "Industry 4.0 technologies assessment: A sustainability perspective", *Int. J. Prod. Econ.,* vol. 229, p. 107776, 2020.
[http://dx.doi.org/10.1016/j.ijpe.2020.107776]

[4]   L. O'Connell, "Size of the global apparel market in 2015 and", Available at: https://www.statista.com/statistics/279735/global-apparel-market-size-by-region/(2017).

[5]   S.P. Yadav, and S. Yadav, "Fusion of Medical Images in Wavelet Domain: A Discrete Mathematical Model", In: *Ingeniería Solidaria.* vol. 14. Universidad Cooperativa de Colombia: UCC, 2018, no. 25, pp. 1-11.
[http://dx.doi.org/10.16925/.v14i0.2236]

[6]   "AI in Fashion Market by Solutions & Services", Available at: https://www.marketsandmarkets.com/Market-Reports/ai-in-fashion-market-144448991(2022).

[7]   S. Golam, and M. Bolesnikov, "A study of digital marketing strategies adopted by chemical industries", Available at: https://www.gbis.ch/index.php/gbis/article/view/36(2021).

[8]   B. Silvestri, "The Future of Fashion: How the Quest for Digitization and the Use of Artificial Intelligence and Extended Reality Will Reshape the Fashion Industry after COVID-19", *ZMJ,* vol. 10, pp. 61-73, 2020.

[9]   Yadav S.P., D. P Mahato, and N. T.D. Linh, *Distributed Artificial Intelligence.* CRC Press., 2020.
[http://dx.doi.org/10.1201/9781003038467]

[10] Y.K. Lee, "Transformation of the innovative and sustainable supply chain with upcoming real-time fashion systems", *Sustainability.,* vol. 13, no. 3, p. 1081, 2021.
[http://dx.doi.org/10.3390/su13031081]

[11] S. Chokshi, and L. Bhattacharya, *Transforming the vision of retail with ai:* Visenze, 2020.

[12] A. Freno, "Practical lessons from developing a large-scale recommender system at zalando", *Proceedings of the Eleventh ACM Conference on Recommender Systems,* pp. 251-259, 2017.
[http://dx.doi.org/10.1145/3109859.3109897]

[13] X. Zou, and W. Wong, "fashion after fashion: A report of ai in fashion", *arXiv preprint arXiv.,* vol. 2105, p. 03050.

[14] S. Park, J. Han, J.Y. Kim, H. Lim, S. Kim, J. Jung, E. Park, S.g. Lee, Y. Lee, and J.Y. Rha, "A deep learning based architecture for personal a.i. fashion stylist services", *The 2nd Artificial Intelligence on Fashion and Textile International Conference (AIFT 2019).*

[15] S. Ren, P. Chi-Leung Hui, and T-m. Jason Choi, "AI-Based Fashion Sales Forecasting Methods in Big Data Era", In: *Artificial Intelligence for Fashion Industry in the Big Data Era.* Springer: Singapore, 2018, pp. 9-26.
[http://dx.doi.org/10.1007/978-981-13-0080-6_2]

[16] R. Schmelzer, "The Fashion Industry Is Getting More Intelligent With AI", Available at: https://www.forbes.com/sites/cognitiveworld/2019/07/16/the-fashion-industry-is-getting-more-intelli-gentwith-ai/(2019).

[17] P. Sheldon, "Augmented reality in retail: Virtually try before you buy", Available at:

https://www.technologyreview.com/s/614616/augmented-reality-in-retail-virtual-try-before-you-buy/

[18]   "How Nike Fit Works", Available at: https://news.nike.com/news/nike-fit-digital-footmeasurem-nt-tool

[19]   "Geneva Environment Network. Environmental Sustainability in the Fashion Industry", Available at: https://www.genevaenvironmentnetwork.org/resources/updates/sustainable-fashion/(2021).

[20]   "United Nations Environment Programme. Fashion's Tiny Hidden Secret", Available at: https://www.unep.org/newsand-stories/story/fashions-tiny-hidden-secret(2019).

[21]   "Ellen MacArthur Foundation. A New Textiles Economy: Redesigning Fashion's Future", Available at: https://ellenmacarthurfoundation.org/a-new-textiles-economy(2017).

[22]   "United Nations Conference on Trade and Development. Report Maps Manufacturing Pollution in Sub-Saharan Africa and South Asia | UNCTA", Available at: https://unctad.org/news/report-map-manufacturing-pollution-in-sub-saharanafrica-and-south-asia(2020).

[23]   "United Nations Environment Programme. Putting the Brakes on Fast Fashion. UN Environment", Available at: https://www.unep.org/news-and-stories/story/putting-brakes-fast-fashion(2018).

[24]   "Environmental Sustainability in the Fashion Industry", Available at: https://www.genevaenvironmen tnetwork.org/resources/updates/sustainable-fashion/(2021).

[25]   "United Nations Environment Programme. Fashion's Tiny Hidden Secret", Available at: https://www.unep.org/newsand-stories/story/fashions-tiny-hidden-secret(2019).

[26]   "Ellen MacArthur Foundation. A New Textiles Economy: Redesigning Fashion's Future", Available at: https://ellenmacarthurfoundation.org/a-new-textiles-economy(2017).

[27]   "United Nations Conference on Trade and Development. Report Maps Manufacturing Pollution in Sub-Saharan Africa and South Asia | UNCTAD", Available at: https://unctad.org/news/report-map-manufacturing-pollution-in-sub-saharanafrica-and-south-asia(2020).

[28]   "United Nations Environment Programme. Putting the Brakes on Fast Fashion. UN Environment", Available at: https://www.unep.org/news-and-stories/story/putting-brakes-fast-fashion(2018).

[29]   M. V. Agarwal, "Green economy research paper", *The International journal of analytical and experimental modal analysis.,* vol. XII, no. IV, 2020.

[30]   V. Jacometti, "Circular Economy and Waste in the Fashion Industry", Available at: www.mdpi.com/journals/laws(2019).
[http://dx.doi.org/10.3390/laws8040027]

[31]   P. Söderholm, "The green economy transition: the challenges of technological change for sustainability", *Sustainable Earth,* vol. 3, no. 1, p. 6, 2020.
[http://dx.doi.org/10.1186/s42055-020-00029-y]

SUBJECT INDEX

A

Ability, sensory-motor 143
Activities 15, 28, 97, 137, 153
  economic 153
  immoral 15
  industrial 97
  nursing 28
  predicting criminal 137
Adenocarcinoma 31
Agent communication 2, 6
  languages (ACLs) 2
  protocol 6
Agents 1, 2, 3, 4, 7, 9, 10, 11, 12, 98, 101, 102, 104, 105, 108, 109, 111, 113
  virtual 98
AI-based 27, 96, 104, 118
  systems 96, 104
  technologies 27, 118
AI-enabled 27, 28, 137, 139, 152, 157, 158
  human brains 27
  hybrid tools 158
  machines 137
  processes 28, 157
  smart cities 139
  technology 152
  tools 158
AIoT 94, 98
  implementation in GSCM 94
  technology 98
AI-powered surveillance system 137
Algorithms 20, 26, 27, 30, 31, 32, 33, 93, 104, 109, 110, 112, 113
  genetic 30, 93
  for machine learning and reasoning 20
Animal bowel anastomosis 33
Apple's Siri 135
AR 48, 49, 56, 57, 61, 86, 144
  and VR technologies 144
  application in education 57
  technology 48, 49, 56, 57, 61, 86

Architecture 39, 49, 80, 95, 98, 101, 109, 110, 113
  computer microchip 98
  intelligent-based decision support 109
Artificial 26, 32, 33, 135
  general intelligence (AGI) 135
  neural networks (ANN) 26, 32, 33
Artificial intelligence 37, 95, 96
  based technology 95
  powered devices 37
  systems 95
  technology 96
AR/VR technology 145
ATMEL microcontroller chip 122
Audio video recording 79
Automated 117, 135
  stock trading 135
  watering system 117
Augmented reality 35, 53
  systems 53
  technologies 35
Automatic robotic surgeon 33
Automation 15, 26, 28, 89, 119, 139
  medical device 28
  processes 119
  robotic process 26
Autonomous vehicles 95

B

Biomarkers, disease 31
Bladder cancer 32
Boltzmann machine 32
Breast cancer 32
Business transitions 90

C

Camera(s) 39, 50, 52, 53, 57, 60, 61, 79, 144
  mobile 50
  sensors 144
Cancer 29, 31, 32, 35

www.ingramcontent.com/pod-product-compliance
Lightning Source LLC
Chambersburg PA
CBHW041704210326
41598CB00007B/520